A FIRST COURSE IN
CHAOTIC DYNAMICAL SYSTEMS
THEORY AND EXPERIMENT

Studies in Nonlinearity

Series Editor: Robert L. Devaney

A FIRST COURSE IN CHAOTIC DYNAMICAL SYSTEMS

THEORY AND EXPERIMENT

Robert L. Devaney

Boston University

Addison-Wesley Publishing Company, Inc.
The Advanced Book Program

Reading, Massachusetts   Menlo Park, California   New York
Don Mills, Ontario   Wokingham, England   Amsterdam   Bonn
Sydney   Singapore   Tokyo   Madrid   San Juan
Paris   Seoul   Milan   Mexico City   Taipei

Mathematics and Physics Editor: *Barbara Holland*
Production Manager: *Pam Suwinsky*
Production Assistant: *Karl Matsumoto*
Editorial Assistant: *Diana Tejo*
Cover Designer: *Nancy Brescia*

Library of Congress Cataloging-in-Publication Data

Devaney, Robert L., 1948–
    A first course in chaotic dynamical systems: theory
and experiment / Robert L. Devaney.
    p. cm. – (Studies in nonlinearity)
    Includes bibliographical references and index.
    1. Differentiable dynamical systems. 2. Chaotic behavior in
systems. I. Title. II. Series: Addison-Wesley studies
in nonlinearity.
    QA614.8.D49 1992   515'.352–dc20   91-38310
    ISBN 0-201-55406-2

2 3 4 5 6 7 8 9 10–MA–96959493
Second printing, March 1993

Preface

   This book is an undergraduate text in dynamical systems. It is aimed at students who have taken at least one year of calculus, but not necessarily any higher level mathematics courses. The book is an outgrowth of a one-semester course taught by the author at Boston University for the past several years. Students in the course ranged from beginning mathematics majors to senior level science and engineering students, from English majors who take mathematics courses "because they are fun" to prospective secondary school mathematics teachers. With the possibility of including computer experimentation, laboratory reports, group projects, together with some wonderfully accessible mathematics, a course in dynamical systems can easily be tailored to such a diverse audience.

   I feel that it is desirable to introduce students to this field at an early stage in their mathematical careers. There are a number of reasons for this. First, the study of dynamics offers mathematicians an opportunity to expose students to contemporary ideas in mathematical research. Many of the ideas and theorems in this book were first discovered within the students' lifetimes; many of the pictures included herein were first viewed within the past decade. To emphasize the contemporary nature of the field, I have included snapshots and brief biographies of a number of individuals who have made recent contributions to the ideas in this book.

   Much of the current interest in dynamics centers around the chaotic behavior that occurs when a simple function is iterated. In this book, the emphasis is on the simplest possible setting in which this occurs, namely iteration of real or complex quadratic polynomials. By dealing mainly with this special case, the material becomes accessible to students who do not have a background in topology or analysis. For example, with only the knowledge of how to multiply complex numbers, students can comprehend the basic mathematical ideas behind such topics as the Julia set or the Mandelbrot set.

A second pedagogical reason to introduce dynamics early in the curriculum is that the course may serve as a bridge between the low–level, often nonrigorous calculus courses and the much more demanding real analysis courses. All too often, students see no connection between the calculus courses that occupy their early years as an undergraduate and the more advanced analysis courses they take later. Dynamics offers students an opportunity to use and build upon their knowledge of calculus and, at the same time, to see in a very concrete setting many of the important topics from basic analysis. I have found that students begin to appreciate the need for abstract metric spaces when they first encounter an object like the space of sequences in symbolic dynamics. They realize the importance of an ϵ-δ definition of continuity when they try to analyze the shift map. Cantor sets become natural objects to study when students see how often they arise in dynamics. Indeed, students who have studied analysis prior to dynamics often remark that they now know why all that abstraction is important!

To accommodate beginning students, this book is structured so that students are gradually introduced to more and more sophisticated ideas from analysis as the chapters unfold. It starts with only a few elementary notions that can be explained using graphical methods or differential calculus. Proofs are introduced slowly at first, and plenty of routine exercises are included. Later come concepts such as dense sets and metric spaces. These concepts arise naturally in the setting of simple dynamical systems, so they can be introduced in a manner that is both concrete and accessible. I feel that this approach is beneficial to those students who do not contemplate future graduate study in mathematics—they see some of the principal ideas of analysis but not in the setting of an intensive course designed for prospective PhDs.

One of the unique aspects of a dynamics course is the possibility of including an experimental component. My students make weekly trips to the computer lab to perform numerical experiments related to the topics covered in class. These experiments range from observations of the rate of convergence to attracting vs. neutral fixed points to a reenactment of Feigenbaum's celebrated discovery of the universality of the period–doubling route to chaos. Students are asked to perform a detailed analysis of the placement of the windows in the orbit diagram as well as an assessment of the meaning of the decorations on the Mandelbrot set. They go to the lab to gather data; they formulate hypotheses and conjectures; they write up their findings in a lengthy lab report. Given the incredible beauty of many of the images the students investigate as well as the open-ended nature of many of the investigations, this portion of the course is always great fun! Moreover, the possibility of combining rigorous mathematics with experimental ideas is a unique opportunity.

Included in the book are a number of sections marked "Experiment." These are the laboratory assignments that my students complete outside of regular class hours. Many of these labs require two weeks to complete, so there are many more experiments included in the text than are possible to complete in a one-semester course. In addition, many of the later labs (especially in the chapter on the Mandelbrot set) demand access to sophisticated computers (certainly including a numeric co-processor).

For this reason, it may be beneficial for instructors to perform some of the labs as classroom demonstrations, asking students to react verbally to what they observe and to discuss what unfolds on the screen. I always use a computer in the classroom to motivate dynamical ideas and to illustrate in "dynamic" fashion what the theorems in the course mean. Of course, the computer does not always give the correct answer. To make students cognizant of this fact and to make sure that they remain suspicious during the course, the first experiments the students perform are entitled "the computer may lie." These experiments should be the first performed, or at least witnessed, by the students.

Software and Solutions Manual

As part of my course, I make extensive use of software for Macintosh computers developed specifically for the course with the assistance of James Georges and Del Johnson. This software, called *A First Course in Chaotic Dynamical Systems Software*, is available from Addison-Wesley and parallels both the experiments and the problems in this book. The software runs on any Macintosh computer with 2 Mb RAM, running System 6.0.5 or higher, and including Color QuickDraw. Site licenses are also available (call Addison-Wesley at 800-447-2226).

The software has been an invaluable aid as both a laboratory and demonstration tool. However, several caveats are in order. The software is not designed as a research tool. Rather, its capabilities are limited by the scope of the experiments and projects in the text. Second, many of the experiments demand a significant amount of computational power or elaborate graphics such as those found on the Macintosh II series of computers. Run times on computers without mathematics coprocessors may be unreasonably long.

On the other hand, the software has been designed so that no prior computer experience on the part of the user is necessary. Indeed, we have segre gated various labs into separate programs that progress in order of difficulty of use. Users are not confronted by a vast array of menu items or options that do everything but wash the dirty dishes in the sink. Rather, each lab has a specific purpose, and the ease of use allows students to concentrate on the mathematics during each lab, rather than the mechanics of making the software work. This means that the students have a significant mathematical experience in the lab rather than a frustrating bout with over-powerful software.

A privately published solutions manual compiled by Thomas R. Scavo is also available for \$12.50 (\$15.00 outside US and Canada) by writing to the author, Robert L. Devaney, at the Mathematics Department, Boston University, 111 Cummington Street, Boston, MA 02215. The manual contains detailed solutions to approximately 75% of the exercises in the text (not including experiments).

Acknowledgments

It is a pleasure to acknowledge the invaluable assistance of Ed Packel, Bruce Peckham, Mark Snavely, Michèle Taylor, and Benjamin Wells, all of whom read and made many fine comments about the manuscript. I am particularly indebted to Tom Scavo for many excellent suggestions concerning both the manuscript and the software. Many of the color plates in this book were produced at Boston University using a program written by Paul Blanchard, Scott Sutherland, and Gert Vegter. Scott Sutherland also assisted with many of the other figures in the book. Thanks are also due Stefen Fangmeier, Chris Mayberry, Chris Small, Sherry Smith, and Craig Upson for help with the computer graphics images. Jim Georges and Del Johnson worked many long hours putting the software for the course into a very usable and enjoyable format; their excellent design for the user interface has made laboratory assignments quite enjoyable and beneficial for my students. Elwood Devaney verified the mathematical accuracy of the entire text; all errors that remain are due to him. Finally, I must also thank my friends Vincenzo, Gaetano, Wolfgang, Giacomo, Gioacchino, Richard, Giuseppe, and Richard for providing me with many hours of enjoyment while this book was taking shape.

Robert L. Devaney

Boston, Massachusetts

January 1993

Contents

[handwritten margin note: "1.5 hr classes / 3 days to here →"]

A FIRST COURSE IN
CHAOTIC DYNAMICAL SYSTEMS
THEORY AND EXPERIMENT

CHAPTER 1

A Mathematical and Historical Tour

Rather than jump immediately into the mathematics of dynamical systems, we will begin with a brief tour of some of the amazing computer graphics images that arise in this field. One of our goals in this book is to explain what these images mean, how they are generated on the computer, and why they are important in mathematics. We will do none of the mathematics here. For now, you should simply enjoy the images. We hope to convince you, in the succeeding chapters, that the mathematics behind these images is even prettier than the pictures. In the second part of the chapter, we will present a brief history of some of the developments in dynamical systems over the past century. You will see that many of the ideas in dynamics arose fairly recently. Indeed, none of the computer graphics images from the tour had been seen before 1980!

1.1   Images from Dynamical Systems

This book deals with some very interesting, exciting, and beautiful topics in mathematics—topics which, in many cases, have been discovered only in the last decade. The main subject of the book is *dynamical systems*, the branch of mathematics that attempts to understand processes in motion. Such processes occur in all branches of science. For example, the motion of the stars and the galaxies in the heavens is a dynamical system, one that has been studied for centuries by thousands of mathematicians and scientists. The stock market is another system that changes in time, as is the world's weather. The changes chemicals undergo, the rise and fall of populations, and the motion of a simple pendulum are classical examples of dynamical systems in chemistry, biology, and physics. Clearly, dynamical systems abound.

What does a scientist wish to do with a dynamical system? Well, since the system is moving or changing in time, the scientist would like to predict where the system is heading, where it will ultimately go. Will the stock market go up or down? Will it be rainy or sunny tomorrow? Will these two chemicals explode if they are mixed in a test tube?

Clearly, some dynamical systems are predictable, whereas others are not. You know that the sun will rise tomorrow and that, when you add cream to a cup of coffee, the resulting "chemical" reaction will not be an explosion. On the other hand, predicting the weather a month from now or the Dow Jones average a week from now seems impossible. You might argue that the reason for this unpredictability is that there are simply too many variables present in meteorological or economic systems. That is indeed true in these cases, but this is by no means the complete answer. One of the remarkable discoveries of twentieth-century mathematics is that very simple systems, even systems depending on only one variable, may behave just as unpredictably as the stock market, just as wildly as a turbulent waterfall, and just as violently as a hurricane. The culprit, the reason for this unpredictable behavior, has been called "chaos" by mathematicians.

Because chaos has been found to occur in the simplest of systems, scientists may now begin to study unpredictability in its most basic form. It is to be hoped that the study of these simpler systems will eventually allow scientists to find the key to understanding the turbulent behavior of systems involving many variables such as weather or economic systems.

In this book we discuss chaos in these simple settings. We will see that chaos occurs in elementary mathematical objects—objects as familiar as quadratic functions—when they are regarded as dynamical systems. You may feel at this point that you know all there is to know about quadratic functions—after all, they are easy to evaluate and to graph. You can differentiate and integrate them. But the key words here are "dynamical systems." We will treat simple mathematical operations like taking the square root, squaring, or cubing as dynamical systems by repeating the procedure over and over, using the output of the previous operation as the input for the next. This process is called *iteration*. This procedure generates a list of real or complex numbers that are changing as we proceed—this is our dynamical system. Sometimes we will find that, when we input certain numbers into the process, the resulting behavior is completely predictable, while other numbers yield results that are often bizarre and totally unpredictable.

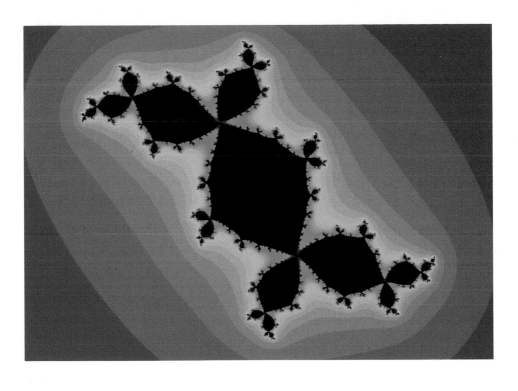

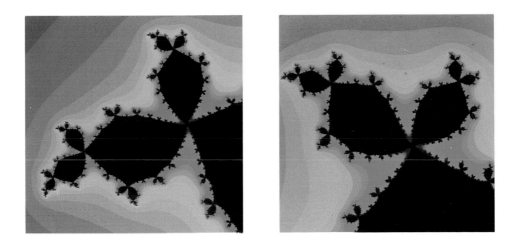

Plates 1, 2, and 3.   Douady's Rabbit and several magnifications

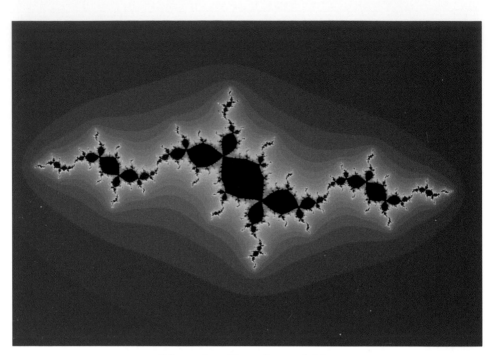

Plate 4.　Dancing rabbits

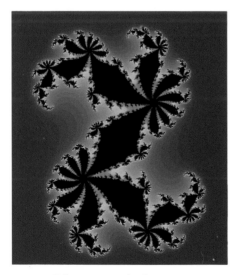

Plate 5.　A dragon

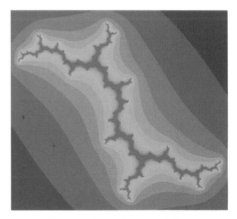

Plate 6.　A dendrite

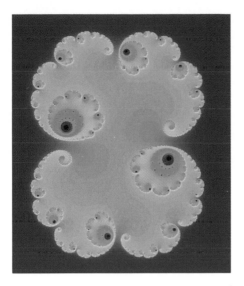

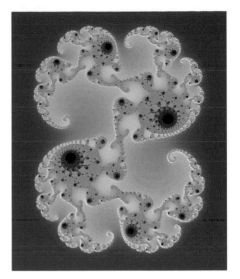

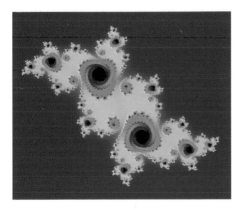

Plates 7–10.

Filled Julia sets for quadratic functions may be Cantor sets.

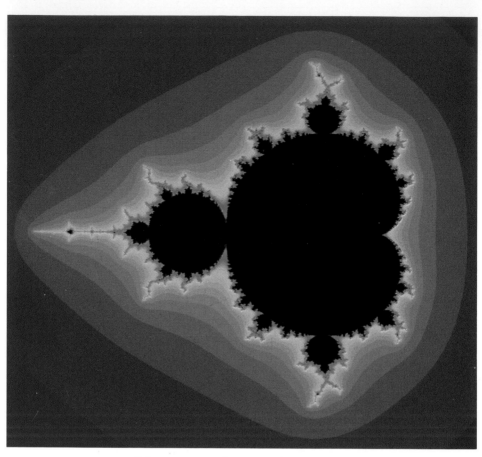

Plate 11.   The Mandelbrot set

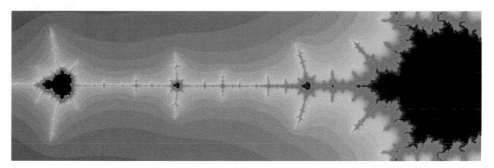

Plate 12.   Tail of the Mandelbrot set

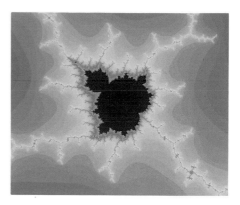

Plates 13, 14.   The period 3 bulb and a magnification

Plates 15, 16.   The period 5 bulb and a magnification

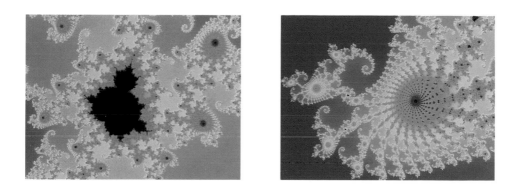

Plates 17–19.   The period 25 bulb and several magnifications

Plates 20–22. Fine detail in the Mandelbrot set

Plate 23.

Antenna attached to the period 3 bulb in the Mandelbrot set

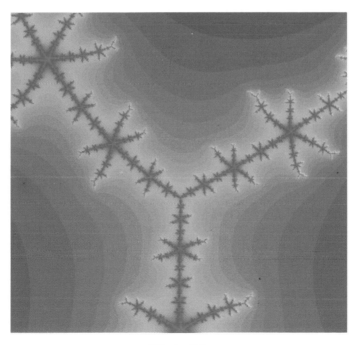

Plate 24.

Julia set corresponding to the junction point in Plate 23

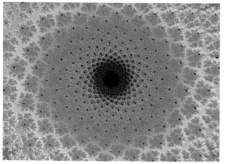 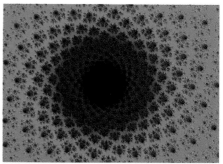

Plates 25–27. Julia set of $(.61 + .81i)\sin(z)$ and several magnifications

Plate 28. Julia set of $(1 + 0.2i)\sin(z)$

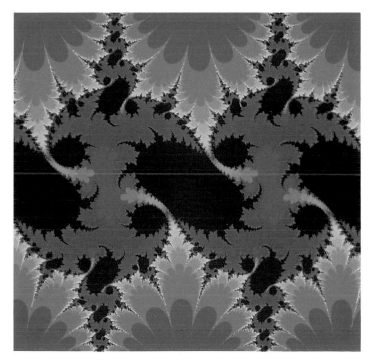

Plate 29. Julia set of $(1 + 0.1i)\sin(z)$

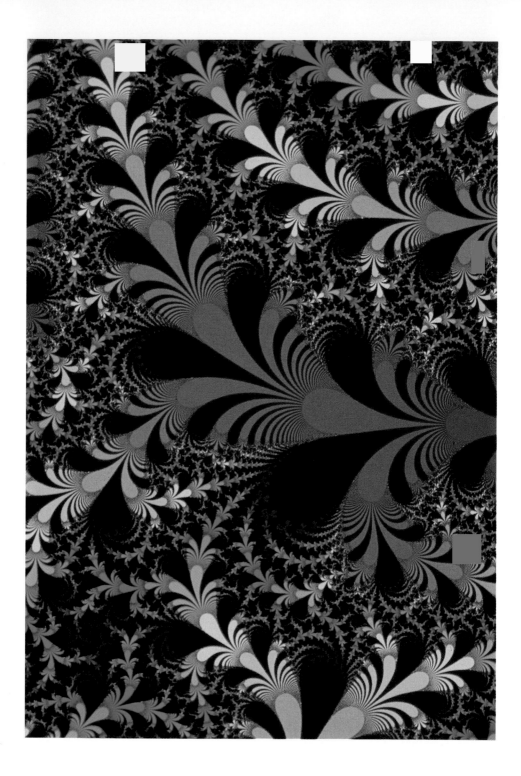

Plate 30.   Julia set of $2\pi i \exp(z)$

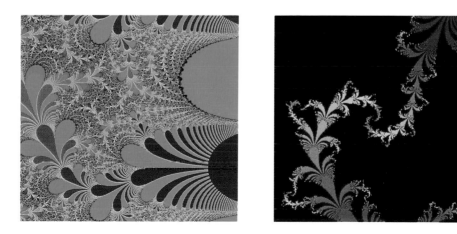

Plates 31–33.   Julia sets of complex exponentials

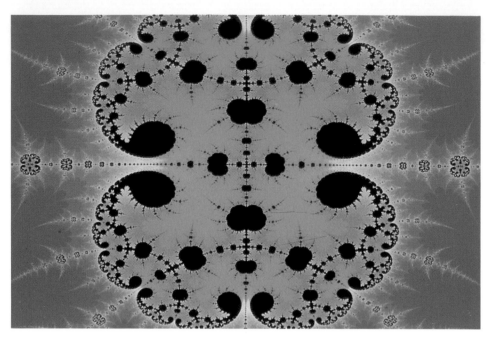

Plate 34.   Julia set of $2.95\cos(z)$

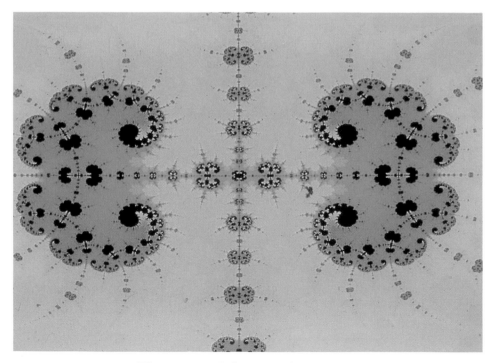

Plate 35.   Julia set of $2.96\cos(z)$

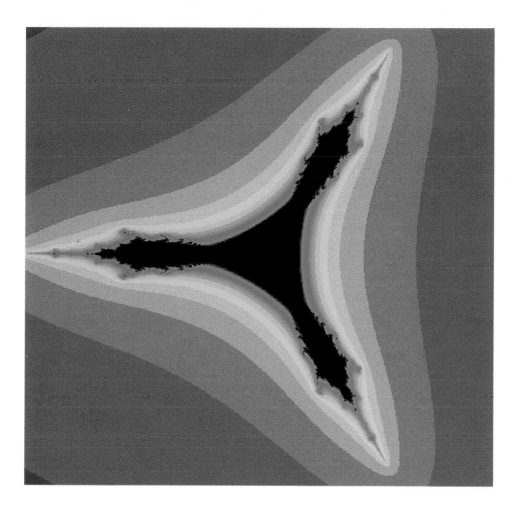

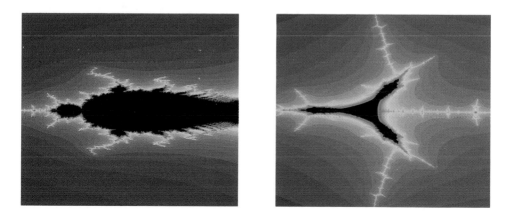

Plates 36–38.   The tricorn and several magnifications

Plate 39. Julia set for Newton's method

For the types of functions we will consider, the set of numbers that yield chaotic or unpredictable behavior in the plane is called the *Julia set* after the French mathematician Gaston Julia, who first formulated many of the properties of these sets in the 1920's. These Julia sets are spectacularly complicated, even for quadratic functions. They are examples of *fractals*. These are sets which, when magnified over and over again, always resemble the original image. The closer you look at a fractal, the more you see exactly the same object. Moreover, fractals naturally have a dimension that is not an integer, not 1, not 2, but often somewhere in between, such as dimension 1.4176, whatever that means! We will discuss these concepts in more detail in Chapter 14.

Here are some examples of the types of images that we will study. In Plate 1 we show the Julia set of the simple mathematical expression $z^2 + c$, where both z and c are complex numbers. In this particular case, $c = -.122 + .745i$. This image is called *Douady's rabbit*, after the French mathematician Adrien Douady whose work we will discuss in Chapter 17. The black region in this image resembles a "fractal rabbit." Everywhere you look, you see a pair of ears. In Plates 2 and 3 we have magnified portions of the rabbit, revealing more and more pairs of ears.

As we will describe later, the black points you see in these pictures are the non-chaotic points. They are points representing values of z that, under iteration of this quadratic function, eventually tend to cycle between three different points in the plane. As a consequence, the dynamical behavior is quite predictable. All of this is by no means apparent right now, but by the time you have read Chapter 16, you will consider this example a good friend. Points that are colored in this picture also behave predictably: They are points that "escape," that tend to infinity under iteration. The colors here simply tell us how quickly a point escapes. The boundary between these two types of behavior—the interface between the escaping and the cycling points—is the Julia set. This is where we will encounter all of the chaotic behavior for this dynamical system.

In Plates 4-10 we have displayed Julia sets for other quadratic functions of the form $z^2 + c$. Each picture corresponds to a different value of c. For example, Plate 6 is a picture of the Julia set for $z^2 + i$. As we see, these Julia sets may assume a remarkable variety of shapes. Sometimes the images contain large black regions as in the case of Douady's rabbit. Other times the Julia set looks like an isolated scatter of points, as in Plates 7–10. Many of these Julia sets are *Cantor sets*. These are very complicated sets that arise often in the study of dynamics. We will begin our study of Cantor sets in Chapter 7 when we introduce the most basic fractal of all, the *Cantor middle-thirds set.*

All these Julia sets correspond to mathematical expressions that are of the form $z^2 + c$. As we see, when c varies, these Julia sets change considerably in shape. How do we understand the totality of all of these shapes, the collection of all possible Julia sets for quadratic functions? The answer is called the *Mandelbrot set.* The Mandelbrot set, as we will see in Chapter 17, is a dictionary, or picture book, of all possible quadratic Julia sets. It is a picture in the c-plane that provides us with a road map of the quadratic Julia Sets. This image, first viewed in 1980 by Benoit Mandelbrot and others, is quite important in dynamics. It completely characterizes the Julia sets of quadratic functions. It has been called one of the most intricate and beautiful objects in mathematics.

Plate 11 shows the full Mandelbrot set. Note that it consists of a basic central cardioid shape, with smaller balls or decorations attached. Plates 13, 15, and 17 are magnifications of some of these decorations. Note how each decoration differs from the others. Buried deep in various regions of the Mandelbrot set, we also find what appear to be small copies of the entire set, as shown in Plates 12, 14, 16, and 18. The set possesses an amazing amount of complexity, as illustrated by Plates 19–22. Nonetheless, each of these small regions has a distinct dynamical meaning, as we will discuss in Chapter 17.

In Plate 23 we have magnified a small area near the "antenna" in the portion of the Mandelbrot set shown in Plate 14. Note that there is a junction point where this antenna seems to branch. In Plate 24 we have displayed a portion of the Julia set for the c-value corresponding to this junction point. Note the remarkable similarity of these two images. This is by no means an accident. As we will see in some of the experiments in Chapter 17, there is an amazing resemblance between certain areas of the Mandelbrot set and the corresponding Julia sets.

In this book we will also investigate the chaotic behavior of many other functions. For example, in Plates 25–27 we have displayed the Julia set of $(0.61 + 0.81i)\sin z$ and several magnifications. Plates 28 and 29 give other examples of Julia sets for functions of the form $c\sin z$. If we investigate exponential functions, we find Julia sets that look quite different as, for example, those in Plates 30-33. Plates 34-35 depict the Julia sets for several cosine functions, while the images in Plates 36-38 are called the *tricorn*, a very different geometric object arising in the study of quadratics. Finally, in Plate 39, we have included a Julia set for Newton's method. This iterative process, familiar from elementary calculus, surprisingly leads to considerable chaotic behavior, as we will see in Chapter 13.

The images in this mathematical tour show quite clearly the great beauty of mathematical dynamical systems theory. But what do these pictures mean and why are they important? These are questions that we will answer in the remainder of this book.

1.2   A Brief History of Dynamics

Dynamical systems has a long and distinguished history as a branch of mathematics. Beginning with the fundamental work of Isaac Newton, differential equations became the principal mathematical technique for describing processes that evolve continuously in time. In the eighteenth and nineteenth centuries, mathematicians devised numerous techniques for solving differential equations explicitly. These methods included Laplace transforms, power series solutions, variation of parameters, linear algebraic methods, and many other techniques familiar from the basic undergraduate course in ordinary differential equations.

There was one major flaw in this development. Virtually all of the analytic techniques for solving differential equations worked mainly for linear differential equations. Nonlinear differential equations proved much more difficult to solve. Unfortunately, many of the most important processes in nature are inherently nonlinear.

An example of this is provided by Newton's original motivation for developing calculus and differential equations. Newton's laws enable us to write down the equations that describe the motion of the planets in the solar system, among many other important physical phenomena. Known as the n-body problem, these laws give us a differential equation whose solution describes the motion of n "point masses" moving in space subject only to their own mutual gravitational attraction. If we know the initial positions and velocities of these masses, then all we have to do is solve Newton's differential equation to be able to predict where and how these masses will move in the future.

This turns out to be a formidable task. If there are only one or two planets, then these equations may be solved explicitly, as is often done in a freshman or sophomore calculus or physics class. For three or more masses, the problem today remains completely unsolved, despite the efforts of countless mathematicians during the past three centuries. It is true that numerical solutions of differential equations by computers have allowed us to approximate the behavior of the actual solutions in many cases, but there are still regimes in the n-body problem where the solutions are so complicated or chaotic that they defy even numerical computation.

Although the explicit solution of nonlinear ordinary differential equations has proved elusive, there have been three landmark events over the past century that have revolutionized the way we study dynamical systems. Perhaps the most important event occurred in 1890. King Oscar II of Sweden announced a prize for the first mathematician who could solve the n-body problem and thereby prove the stability of the solar system. Needless to say, nobody solved the original problem, but the great French mathematician Henri Poincaré came closest. In a beautiful and far-reaching paper, Poincaré totally revamped the way we tackle nonlinear ordinary differential equations. Instead of searching for explicit solutions of these equations, Poincaré advocated working qualitatively, using topological and geometric techniques, to uncover the global structure of all solutions. To him, a knowledge of all possible behaviors of the system under investigation was much more important than the rather specialized study of individual solutions.

Poincaré's prize winning paper contained a major new insight into the behavior of solutions of differential equations. In describing these solutions, mathematicians had previously made the tacit assumption that what we now know as stable and unstable manifolds always match up. Poincaré questioned this assumption. He worked long and hard to show that this was always the case, but he could not produce a proof. He eventually concluded that the stable and unstable manifolds might not match up and could actually cross at an angle. When he finally admitted this possibility, Poincaré saw that this would cause solutions to behave in a much more complicated fashion than anyone had previously imagined. Poincaré had discovered what we now call *chaos*. Years later, after many attempts to understand the chaotic behavior of differential equations, he threw up his hands in defeat and wondered if anyone would ever understand the complexity he was finding. Thus, "chaos theory," as it is now called, really dates back over 100 years to the work of Henri Poincaré.

Poincaré's achievements in mathematics went well beyond the field of dynamical systems. His advocacy of topological and geometric techniques opened up whole new subjects in mathematics. In fact, building on his ideas, mathematicians turned their attention away from dynamical systems and toward these related fields in the ensuing decades. Areas of mathematics such as algebraic and differential topology were born and flourished in the twentieth century. But nobody could handle the chaotic behavior that Poincaré had observed, so the study of dynamics languished.

There were two notable exceptions to this. One was the work of the French mathematicians Pierre Fatou and Gaston Julia in the 1920's on the dynamics of complex analytic maps. They too saw chaotic behavior, this time on what we now call the *Julia set*. Indeed, they realized how tremendously intricate these Julia sets could be, but they had no computer graphics available to see these sets, and as a consequence, this work also stopped in the 1930's.

At the same time, the American mathematician G. D. Birkhoff adopted Poincaré's qualitative point of view on dynamics. He advocated the study of iterative processes as a simpler way of understanding the dynamical behavior of differential equations, a viewpoint that we will adopt in this book.

The second major development in dynamical systems occurred in the 1960's. The American mathematician Stephen Smale reconsidered Poincaré's crossing stable and unstable manifolds from the point of view of iteration and showed by an example that the chaotic behavior that baffled his predecessors could indeed be understood and analyzed completely. The technique he used to analyze this is called *symbolic dynamics* and will be a major tool for us in this book. At the same time, the American meteorologist E. N. Lorenz, using a very crude computer, discovered that very simple differential equations could exhibit the type of chaos that Poincaré observed. Lorenz, who actually had been a Ph.D. student of Birkhoff, went on to observe that his simple meteorological models exhibited what is now called *sensitive dependence on initial conditions*. For him, this meant that long-range weather forecasting was all but impossible and showed that the mathematical topic of chaos was important in other areas of science.

This led to a tremendous flurry of activity in nonlinear dynamics in the 1970's. The ecologist Robert May found that very simple iterative processes that arise in mathematical biology could produce incredibly complex and chaotic behavior. The physicist Mitchell Feigenbaum, building on Smale's earlier work, noticed that, despite the complexity of chaotic behavior, there was some semblance of order in the way systems became chaotic. Physicists Harry Swinney and Jerry Gollub showed that these mathematical developments could actually be observed in physical applications, notably in turbulent fluid flow. More recently, other systems, such as the motion of the planet Pluto or the beat of the human heart, have been shown to exhibit similar chaotic patterns. In mathematics, meanwhile, new techniques were developed to help understand chaos. John Guckenheimer and Robert

F. Williams employed the theory of strange attractors to explain the phenomenon that Lorenz had observed a decade earlier. And tools such as the Schwarzian derivative, symbolic dynamics, and bifurcation theory—all topics we will discuss in this book—were shown to play an important role in understanding the behavior of dynamical systems.

The third and most recent major development in dynamical systems was the availability of high-speed computing and, particularly, computer graphics. Foremost among the computer-generated results was Mandelbrot's discovery in 1980 of what is now called the *Mandelbrot set*. This beautiful image immediately reawakened interest in the old work of Julia and Fatou. Using the computer images as a guide, mathematicians such as Adrien Douady, John Hubbard, and Dennis Sullivan greatly advanced the classical theory. Other computer graphics images such as the orbit diagram and the Lorenz attractor generated considerable interest among mathematicians and led to further advances.

One of the most interesting side effects of the availability of high speed computing and computer graphics has been the development of an experimental component in the study of dynamical systems. Whereas the old masters had to rely solely on their imagination and their intellect, now mathematicians have an invaluable additional resource to investigate dynamics: the computer. This tool has opened up whole new vistas for dynamicists, some of which we will sample in this book. In a series of sections called "Experiments," you will have a chance to rediscover some of these wonderful facts yourself.

CHAPTER 2

Examples of Dynamical Systems

Our goal in this book is to describe the beautiful mathematical subject known as dynamical systems theory. We will, for the most part, concentrate on the mathematics itself rather than the applications of the subject. However, for motivation, we will briefly describe in this chapter four different examples of dynamical systems that arise in practice.

2.1   An Example from Finance

Consider the following situation. Suppose we deposit $1000 in a bank at 10 percent interest. We ask the question: if we leave this money untouched for n years, how much money will we have in our account at the end of this period? For simplicity, we assume that the 10 percent interest is added to our account once each year at the end of the year.

This is one of the simplest examples of an *iterative process* or *dynamical system*. Let's denote the amount we have in the bank at the end of the nth year by A_n. Our problem is to determine A_n for some given number of years n. We know that A_0, our initial deposit, is $1000. After 1 year we add 10 percent to this amount to get our new balance. That is,

$$A_1 = A_0 + 0.1A_0 = 1.1A_0.$$

In our specific case, $A_1 = \$1100$. At the end of the second year, we perform the same operation

$$A_2 = A_1 + 0.1A_1 = 1.1A_1,$$

so that $A_2 = \$1210$. Continuing,

$$A_3 = 1.1A_2$$
$$A_4 = 1.1A_3$$
$$\vdots$$
$$A_n = 1.1A_{n-1}.$$

Thus we can recursively determine the amount A_n once we know the previous year's balance.

The equation $A_n = 1.1A_{n-1}$ is an example of a (first-order) *difference equation*. In such an equation, we use information from the previous year (or other fixed time interval) to determine the current information, then we use this information to determine next year's amount, and so forth.*

We solve this difference equation by the process of iteration. The iterative process involved is multiplication by 1.1. That is, if we define the function $F(x) = 1.1x$, then our savings balances are determined by repeatedly applying this function:

$$A_1 = F(A_0)$$
$$A_2 = F(A_1)$$
$$A_3 = F(A_2)$$

and so forth. Note that we may also write

$$A_2 = F(F(A_0)) = F \circ F(A_0)$$
$$A_3 = F(F(F(A_0))) = F \circ F \circ F(A_0)$$

to clearly indicate that we compose F with itself repeatedly to obtain the successive balances.

Since $F(x) = 1.1x$, we have

$$F(F(x)) = (1.1)^2 x$$
$$F(F(F(x))) = (1.1)^3 x,$$

and, in general, the nth iteration of the function yields

$$\underbrace{F \circ \cdots \circ F}_{n \text{ times}}(x) = (1.1)^n x.$$

---

*   In higher order difference equations, which we will not discuss in this book, we would need information from several prior years to determine the current value.

So to find A_n, we merely compute $(1.1)^n$ and multiply by A_0. For example, using a calculator or computer, you may easily check that $A_{10} = \$2593.74$ and $A_{50} = \$117,390.85$.

This example is quite simple: the iterations we will encounter will in general yield much more complicated results.

2.2   An Example from Ecology

Here is another difference equation which is essentially the same as our savings account example. Suppose we wish to predict the behavior of the population of a certain species which grows or declines as generations pass. Let's denote the population alive at generation n by P_n. So our question is: can we predict what will happen to P_n as n gets large? Will P_n tend to zero so that the species becomes extinct? Or will P_n grow without bound so that we experience a population explosion?

There are many mathematical models to predict the behavior of populations. By far the simplest (and most naive) is the exponential growth model. In this model we assume that the population in the succeeding generation is directly proportional to the population in the current generation. This translates to mathematics as another difference equation

$$P_{n+1} = rP_n$$

where r is some constant determined by ecological conditions. Thus, given the initial population P_0, we can recursively determine the population in the succeeding generations:

$$P_1 = rP_0$$
$$P_2 = rP_1 = r^2 P_0$$
$$P_3 = rP_2 = r^3 P_0$$
$$\vdots$$
$$P_n = rP_{n-1} = r^n P_0.$$

As before, we determine the behavior of the population via iteration. In this case, the function we iterate is $F(x) = rx$. So

$$P_n = \underbrace{F \circ \cdots \circ F}_{n \text{ times}}(P_0) = r^n P_0.$$

Note that the ultimate fate of the population depends on r. If $r > 1$, then r^n tends to infinity with n, so we have unchecked population growth. If $r < 1$, r^n tends to zero, so the species becomes extinct. Finally, if $r = 1$, $P_n = P_0$, so there is never any change in the population. Thus, we can achieve our goal of determining the fate of the species for any r. Of course, this simplified model is highly unrealistic in that real-life populations behave in a much more complicated fashion. For example, populations can never tend to infinity. To remedy this, we will add one assumption to our model that will take into account the possibility of overcrowding.

Specifically, we will discuss what ecologists call the *logistic model of population growth*. In this model, we assume at the outset that there is some absolute maximum population that can be supported by the environment. If the population ever reaches this number, then we have disastrous overcrowding—food supply becomes critically short—and the species immediately dies out. To keep the numbers manageable, let's assume that P_n now represents the fraction of this maximal population alive at generation n, so that $0 \le P_n \le 1$. The logistic model is then

$$P_{n+1} = \lambda P_n(1 - P_n).$$

As before, λ is a constant that depends on ecological conditions. For reasons that will become apparent later, we will always assume that $0 < \lambda \le 4$.

Note that, in the absence of the $1 - P_n$ factor, we are left with the previous exponential growth model. If $P_n = 0$ (no individuals present), then $P_{n+1} = 0$ as well, as we would expect. If $P_n = 1$, then $P_{n+1} = 0$ as we have assumed.

Thus, to understand the growth and decline of the population under this model, we must iterate the *logistic function* $F_\lambda(x) = \lambda x(1 - x)$. Unlike the previous examples, this function is quadratic rather than linear. We will see that this simple change gives rise to a very rich mathematical theory. Indeed, the behavior of this function under iteration is still far from being completely understood. We will spend most of this book analyzing the dynamical behavior of this and other similar functions.

2.3   Finding Roots and Solving Equations

How do you find $\sqrt{5}$ exactly? Believe it or not, the simplest method dates back to the time of the Babylonians and involves a simple iteration. We will make an initial guess x_0 for $\sqrt{5}$. We assume that x_0 is positive. Now, chances are that $x_0 \ne \sqrt{5}$, so we will use this guess to produce a new and better guess x_1.

Here is the procedure. If $x_0 \neq \sqrt{5}$, then we either have $x_0 < \sqrt{5}$ or $x_0 > \sqrt{5}$. In the former case, we have

$$\sqrt{5}x_0 < 5$$

$$\sqrt{5} < \frac{5}{x_0}$$

for $x_0 \neq 0$. On the other hand, if $x_0 > \sqrt{5}$, then

$$\sqrt{5} > \frac{5}{x_0}.$$

Thus we have either

$$x_0 < \sqrt{5} < \frac{5}{x_0}$$

or

$$\frac{5}{x_0} < \sqrt{5} < x_0.$$

Therefore, if we take the average of x_0 and $5/x_0$, namely

$$x_1 = \frac{1}{2}\left(x_0 + \frac{5}{x_0}\right),$$

the resulting value will lie midway between x_0 and $5/x_0$ and so will, hopefully, be a better approximation to $\sqrt{5}$. So we use this value as our next "guess" for $\sqrt{5}$. Continuing, we form the successive averages

$$x_2 = \frac{1}{2}\left(x_1 + \frac{5}{x_1}\right)$$

$$x_3 = \frac{1}{2}\left(x_2 + \frac{5}{x_2}\right)$$

$$\vdots$$

Intuitively, the sequence of numbers $x_0, x_1, x_2, \ldots$ should eventually approach $\sqrt{5}$.

Let's see how this works in practice. Suppose we make the (somewhat silly) initial guess $x_0 = 1$ for $\sqrt{5}$. Then we have

$$x_1 = \frac{1}{2}(1+5) = 3$$

$$x_2 = \frac{1}{2}\left(3 + \frac{5}{3}\right) = \frac{7}{3} = 2.333\ldots$$

$$x_3 = \frac{1}{2}(\frac{7}{3} + \frac{15}{7}) = \frac{47}{21} = 2.238095\ldots$$

$$x_4 = 2.236068\ldots$$

$$x_5 = 2.236067\ldots$$

$$x_6 = 2.236067\ldots,$$

and we see that, very quickly, this sequence tends to the correct answer, as $\sqrt{5} = 2.236067\ldots$.

Clearly, to find other square roots, we need only replace the 5 in the formula for our initial guess. Thus, to "find" $\sqrt{9}$, we choose some initial guess, say $x_0 = 2$, and then compute

$$x_1 = \frac{1}{2}\left(2 + \frac{9}{2}\right) = 3.25$$

$$x_2 = \frac{1}{2}\left(x_1 + \frac{9}{x_1}\right) = 3.0096\ldots$$

$$x_3 = 3.000015\ldots$$

$$x_4 = 3.000000\ldots$$

$$x_5 = 3.000000\ldots,$$

and we see that this sequence quickly converges to $\sqrt{9} = 3$.

A related question that arises in all branches of science and mathematics is, How do you solve the equation $F(x) = 0$? For example, finding the square root of 5 is the same as solving the equation $x^2 - 5 = 0$. There are very few functions for which it is possible to write down the solutions to this equation explicitly. Even for polynomials, solving this equation is difficult when the degree is greater than 2, and generally impossible when the degree is 5 or greater. Yet the problem is extremely important, so we seek other possible methods.

One method, familiar from calculus, is Newton's method. This method involves the following procedure. Given a function F whose roots we are trying to find, we construct a new function, the Newton iteration function, given by

$$N(x) = x - \frac{F(x)}{F'(x)}.$$

Then we make an initial guess x_0 for a root. Newton's method is to iterate the function N, successively computing

$$x_1 = N(x_0)$$
$$x_2 = N(x_1) = N(N(x_0))$$
$$x_3 = N(x_2) = N(N(N(x_0)))$$

and so forth. Often, though by no means always, this sequence of iterates converges to a root of F. Note that, in the special case where $F(x) = x^2 - 5$ (whose roots are $\pm\sqrt{5}$), the Newton iteration is simply

$$N(x) = x - \frac{x^2 - 5}{2x} = \frac{1}{2}\left(x + \frac{5}{x}\right),$$

as was discussed above.

We will not take the time now to discuss why this happens or even where the Newton iteration function comes from. Rather, we will devote all of Chapter 13 to this subject.

With the advent of accessible high-speed computation, iterative algorithms such as Newton's method are becoming an important and widespread area of interest in mathematics. No longer are such algorithms primarily of theoretical interest!

2.4   Differential Equations

Differential equations are also examples of dynamical systems. Unlike iterative processes where time is measured in discrete intervals such as years or generations, differential equations are examples of *continuous* dynamical systems wherein time is a continuous variable. Ever since the time of Newton, these types of systems have been of paramount importance.

Recently, however, *discrete* dynamical systems have also received considerable attention. This does not mean that continuous systems have declined in importance. Rather, mathematicians study discrete systems with an eye toward applying their results to the more difficult continuous case.

There are a number of ways that iterative processes enter the arena of differential equations. For example, a solution of a differential equation is a continuous function of time. If we look at this solution at discrete time intervals, say at times $t = 0, 1, 2, \ldots$, then we are really considering an iterative process as described above.

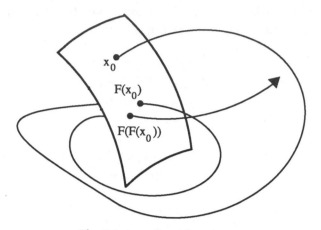

Fig. 2.1 A surface of section.

Most often, differential equations are impossible to solve explicitly. We must turn to the computer to generate numerical solutions. The numerical methods used to solve these equations are often iterative processes such as the Runge-Kutta method.

A final way in which iteration arises in the study of differential equations occurs when a surface of section can be found. Suppose that we have a first-order differential equation in three dimensions whose independent variable is time. Then the solutions we seek are curves in space parameterized by time. Often, though not always, these curves intersect a given surface in space over and over again as depicted in Figure 2.1. When this occurs, the study of solutions of the equation reduces to the study of an iterative process on the surface. Starting with any point on the surface x_0, we follow the solution through x_0 until it first reintersects the surface. Call this point of first return $F(x_0)$. The function F is called a *first return map* and the surface is a *surface of section*. Continuing to follow the solution curve, its next point of intersection is $F(F(x_0))$, so we see that determining the successive points of intersection is really a problem of iterating F.

Iteration of the first return map does not tell us all there is to know about the solutions of the differential equation, but it does give us a lot of qualitative information. In particular, we do not know the exact location of a solution at each moment of time, but we do gain long-term information about the behavior of the solution.

CHAPTER 3

Orbits

3.1   Iteration

As we saw in the previous chapter, there are many kinds of problems in science and mathematics that involve *iteration*. Iteration means to repeat a process over and over. In dynamics, the process that is repeated is the application of a function.

In this and the next few chapters we will consider only functions of one variable as encountered in elementary calculus. We will spend quite a bit of time discussing the *quadratic* functions $Q_c(x) = x^2 + c$ where $c \in \mathbf{R}$ (a real number) is a constant. Other functions that will arise often are the *logistic* functions $F_\lambda(x) = \lambda x(1 - x)$, the *exponentials* $E_\lambda(x) = \lambda e^x$, and the *sine* functions $S_\mu(x) = \mu \sin x$. Here, λ and μ are constants. The constants c, μ, and λ are called *parameters*. One of the important questions we will address later is how the dynamics of these functions change as these parameters are varied.

To iterate a function means to evaluate the function over and over, using the output of the previous application as the input for the next. This is the same process as typing a number into a scientific calculator, then repeatedly striking one of the function keys such as "sin" or "cos." Mathematically, this is the process of repeatedly composing the function with itself.

We write this as follows. For a function F, $F^2(x)$ is the second iterate of F, namely $F(F(x))$, $F^3(x)$ is the third iterate $F(F(F(x)))$, and, in general, $F^n(x)$ is the n-fold composition of F with itself. For example, if $F(x) = x^2 + 1$, then

$$F^2(x) = (x^2 + 1)^2 + 1$$

$$F^3(x) = ((x^2 + 1)^2 + 1)^2 + 1.$$

Similarly, if $F(x) = \sqrt{x}$, then

$$F^2(x) = \sqrt{\sqrt{x}}$$

$$F^3(x) = \sqrt{\sqrt{\sqrt{x}}}.$$

It is important to realize that $F^n(x)$ does not mean raise $F(x)$ to the nth power (an operation we will never use). Rather, $F^n(x)$ is the nth iterate of F evaluated at x.

3.2   Orbits

Given $x_0 \in \mathbf{R}$, we define the *orbit of x_0 under F* to be the sequence of points $x_0, x_1 = F(x_0), x_2 = F^2(x_0), \ldots, x_n = F^n(x_0), \ldots$. The point x_0 is called the *seed* of the orbit.

For example, if $F(x) = \sqrt{x}$ and $x_0 = 256$, the first few points on the orbit of x_0 are

$$x_0 = 256$$
$$x_1 = \sqrt{256} = 16$$
$$x_2 = \sqrt{16} = 4$$
$$x_3 = \sqrt{4} = 2$$
$$x_4 = \sqrt{2} = 1.41\ldots.$$

As another example, if $S(x) = \sin x$ (where x is given in radians, not degrees), the orbit of $x_0 = 123$ is

$$x_0 = 123$$
$$x_1 = -0.4599\ldots$$
$$x_2 = -0.4438\ldots$$
$$\vdots$$
$$x_{300} = -0.0975\ldots$$
$$x_{301} = -0.0974\ldots$$
$$\vdots$$

Slowly, ever so slowly, the points on this orbit tend to 0.

If $C(x) = \cos x$, then the orbit of $x_0 = 123$ is

$$x_0 = 123$$
$$x_1 = -0.8879\ldots$$
$$x_2 = 0.6309\ldots$$

$$\vdots$$

$$x_{50} = 0.739085$$
$$x_{51} = 0.739085$$
$$x_{52} = 0.739085$$

$$\vdots$$

After only a few iterations, this orbit seems to stop at $0.739085\ldots$.

3.3   Types of Orbits

There are many different kinds of orbits in a typical dynamical system. Undoubtedly the most important kind of orbit is a *fixed point*. A fixed point is a point x_0 that satisfies $F(x_0) = x_0$. Note that $F^2(x_0) = F(F(x_0)) = F(x_0) = x_0$ and, in general, $F^n(x_0) = x_0$. So the orbit of a fixed point is the constant sequence $x_0, x_0, x_0, \ldots$. A fixed point never moves. As its name implies, it is fixed by the function. For example, 0, 1, and -1 are all fixed points for $F(x) = x^3$, while only 0 and 1 are fixed points for $F(x) = x^2$. Fixed points are found by solving the equation

$$F(x) = x.$$

Thus, $F(x) = x^2 - x - 4$ has fixed points at the solutions of

$$x^2 - x - 4 = x,$$

which are $1 \pm \sqrt{5}$, as determined by the quadratic formula. (Remember to bring the right-hand x to the other side!)

Fixed points may also be found geometrically by examining the intersection of the graph with the diagonal line $y = x$. For example, Figure 3.1 shows that the only fixed point of $S(x) = \sin x$ is $x_0 = 0$, since that is the only point of intersection of the graph of S with the diagonal $y = x$. Similarly, $C(x) = \cos x$ has a fixed point at $0.739085\ldots$, as shown in Figure 3.2.

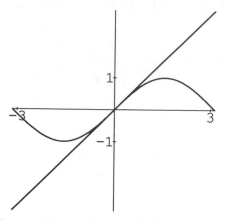

Fig. 3.1. The fixed point of $S(x) = \sin x$ is 0.

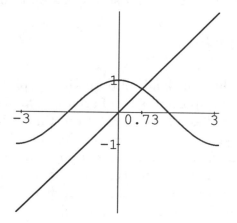

Fig. 3.2 The fixed point of $C(x) = \cos x$ is $0.739085\ldots$.

Another important kind of orbit is the *periodic orbit* or *cycle*. The point x_0 is periodic if $F^n(x_0) = x_0$ for some $n > 0$. The least such n is called the *prime period* of the orbit. Note that if x_0 is periodic with prime period n, then the orbit of x_0 is just a repeating sequence of numbers

$$x_0, F(x_0), \ldots, F^{n-1}(x_0), x_0, F(x_0), \ldots, F^{n-1}(x_0), \ldots.$$

For example, 0 lies on a cycle of prime period 2 for $F(x) = x^2 - 1$, since $F(0) = -1$ and $F(-1) = 0$. Thus the orbit of 0 is simply

$$0, -1, 0, -1, 0, -1, \ldots.$$

We also say that 0 and -1 form a 2-cycle. Similarly, 0 lies on a periodic orbit of prime period 3 or a 3-cycle for $F(x) = -\frac{3}{2}x^2 + \frac{5}{2}x + 1$, since $F(0) =$

$1, F(1) = 2$, and $F(2) = 0$. So the orbit is

$$0, 1, 2, 0, 1, 2, \ldots.$$

We will see much later that the appearance of this seemingly harmless 3-cycle has surprising implications for the dynamics of this function.

In general, it is very difficult to find periodic points exactly. For example, to find cycles of period 5 for $F(x) = x^2 - 2$, we would have to solve the equation

$$F^5(x) - x = 0.$$

This is a polynomial equation of degree $2^5 = 32$. Polynomial equations of such high degree are usually impossible to solve exactly. More generally, to find cycles of period n for this function, we would have to solve a polynomial equation of degree 2^n, clearly an impossible task.

Note that if x_0 has prime period k, then x_0 is also fixed by F^{2k}. Indeed $F^{2k}(x_0) = F^k(F^k(x_0)) = F^k(x_0) = x_0$. Similarly, x_0 is fixed by F^{nk}, so we say that x_0 has period nk for any positive integer n. We reserve the word *prime* period for the case $n = 1$.

Also, if x_0 lies on a periodic orbit of period k, then all points on the orbit of x_0 have period k as well. Indeed, the orbit of x_1 is

$$x_1, x_2, \ldots, x_{k-1}, x_0, x_1, \ldots, x_{k-1}, x_0, x_1 \ldots,$$

which has period k.

A point x_0 is called *eventually fixed* or *eventually periodic* if x_0 itself is not fixed or periodic, but some point on the orbit of x_0 is fixed or periodic. For example, -1 is eventually fixed for $F(x) = x^2$, since $F(-1) = 1$, which is fixed. Similarly, 1 is eventually periodic for $F(x) = x^2 - 1$ since $F(1) = 0$, which lies on a cycle of period 2. The point $\sqrt{2}$ is also eventually periodic for this function, since the orbit is

$$\sqrt{2}, 1, 0, -1, 0, -1, 0, -1, \ldots.$$

In a typical dynamical system, most orbits are not fixed or periodic. For example, for the linear function $T(x) = 2x$, only 0 is a fixed point. All other orbits of T get larger and larger (in absolute value) under iteration since $T^n(x_0) = 2^n x_0$. Indeed, if $x_0 \neq 0$, $|T^n(x_0)|$ tends to infinity as n approaches infinity. We denote this by

$$|T^n(x_0)| \to \infty.$$

The situation is reversed for the linear function $L(x) = \frac{1}{2}x$. For L, only 0 is fixed, but for any $x_0 \neq 0$,

$$L^n(x_0) = \frac{x_0}{2^n}.$$

Since the sequence $\{\frac{1}{2^n}\}$ tends to zero, we have

$$L^n(x_0) \to 0.$$

We say that the orbit of x_0 converges to the fixed point 0.

As another example, consider the squaring function $F(x) = x^2$. If $|x_0| < 1$, it is easy to check that $F^n(x_0) \to 0$. For example, if $x_0 = 0.1$, then the orbit of x_0 is

$$0.1, 0.01, 0.0001, \ldots, 10^{-2^n}, \ldots,$$

which clearly tends to zero.

This example points out one danger in using a computer or calculator to iterate functions. If we input 0.1 into a scientific calculator and then depress the "x^2" key repeatedly, we are computing the orbit of 0.1 under the squaring function. However, after a few iterations, the calculator display will read $0.000\ldots$. This, of course, is false. We know that the nth point on the orbit of 0.1 is 10^{-2^n}, which is non-zero. The calculator, however, can store numbers up to only a finite number of decimal places. Hence 10^{-2^n} is represented as 0 for large enough n. We will always have to keep this in mind in later sections when we compute orbits numerically. You should be forewarned that this is not the only way that a computer will lie to us!

3.4   Other Orbits

The simple orbits we have discussed so far—fixed, periodic, and eventually periodic orbits as well as orbits that tend to a specific limit—might tempt you to think that dynamical systems typically have very simple behavior. Actually, nothing could be further from the truth. One of the major discoveries in mathematics over the past twenty-five years is that many simple functions—such as quadratic functions of a real variable—may have many orbits of incredible complexity.

We give here two very simple examples. Consider the quadratic function $F(x) = x^2 - 2$. The orbit of 0 is very simple—it is eventually fixed:

$$0, -2, 2, 2, 2, \ldots.$$

But consider the orbits of nearby points. In Table 1 we have listed the first few points on the orbits of three nearby points. Notice what happens: after a very few iterations, these orbits are far from being fixed. In fact, they seem to wander almost randomly about the interval from -2 to 2. This is our first view of chaotic behavior, one of the main subjects of this book.

n	x=0	x=0.1	x=0.01	x=0.001
0	0	0.1	0.01	0.001
1	-2	-1.99	-1.999	-1.999
2	2	1.960	1.999	1.999
3	2	1.842	1.998	1.999
4	2	1.393	1.993	1.999
5	2	-0.597	1.974	1.999
6	2	-1.996	1.898	1.999
7	2	1.986	1.604	1.996
8	2	1.943	0.573	1.984
9	2	1.776	-1.671	1.938
10	2	1.154	0.793	1.755
11	2	-0.669	-1.370	1.081
12	2	-1.552	-0.122	-0.832
13	2	0.410	-1.985	-1.307
$\vdots$	$\vdots$	$\vdots$	$\vdots$	$\vdots$
95	2	1.999	1.548	1.048
96	2	1.999	0.398	-0.902
97	2	1.996	-1.841	-1.186
98	2	1.986	1.391	-0.593
99	2	1.946	-0.065	-1.648
100	2	1.789	-1.995	0.710

Table 1 Several orbits of $x^2 - 2$.

Here is another view of one of these orbits. In Figure 3.3 we display a histogram of the orbit of 0.1 under $x^2 - 2$. To compute this image, we subdivided the interval $-2 \leq x \leq 2$ into 400 subintervals of equal width 0.01. We then computed 20,000 points on the orbit of 0.1 and marked off one point over each subinterval each time the orbit entered that interval. Note how the orbit has distributed itself not quite evenly over the interval $-2 \leq x \leq 2$. It has, however, visited each subinterval a substantial number of times. This, too, is one of the ingredients of chaotic behavior to which we will return later.

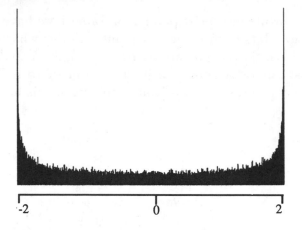

Fig. 3.3 A histogram of the orbit of 0.1 under $x^2 - 2$.

3.5   The Doubling Function

Our next example is one to which we will return often in this book—the doubling function. This function has domain the half-open, half-closed unit interval $0 \leq x < 1$ which we denote by $[0, 1)$. The doubling function D is defined by

$$D(x) = \begin{cases} 2x & 0 \leq x < 1/2 \\ 2x - 1 & 1/2 \leq x < 1. \end{cases}$$

The graph of D is displayed in Figure 3.4. Note that $D \colon [0, 1) \to [0, 1)$.

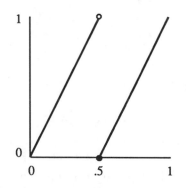

Fig. 3.4 The graph of the doubling function.

We may define D more succinctly as

$$D(x) = 2x \mod 1.$$

This means that $D(x)$ is the fractional part of $2x$. For example, $D(0.3) = 0.6$, but $D(0.6) = 1.2 - 1 = 0.2$, and $D(0.9) = 0.8$.

Note that D has lots of cycles. For example, 0 is a fixed point. The points 1/3 and 2/3 lie on a 2-cycle. The point 1/5 lies on a 4-cycle:

$$\frac{1}{5}, \frac{2}{5}, \frac{4}{5}, \frac{3}{5}, \frac{1}{5}, \frac{2}{5}, \ldots$$

And 1/9 lies on a 6-cycle:

$$\frac{1}{9}, \frac{2}{9}, \frac{4}{9}, \frac{8}{9}, \frac{7}{9}, \frac{5}{9}, \frac{1}{9}, \ldots$$

It turns out that there are many other types of orbits for this function, including some that are not periodic, eventually periodic, or convergent to a cycle. Indeed, as we will see later, *most* orbits are of none of these types. However, on a typical digital computer or scientific calculator we never (well, okay, almost never) see any of these orbits. This is the content of our first experiment.

3.6   Experiment: The Computer May Lie

Goal: The aim of this experiment is to compute a number of orbits for several different functions and record the results. We will see as we proceed that the results generated by the computer can often be misleading or just plain wrong!

Procedure: Consider the following three functions.
1. $F(x) = x^2 - 2$.
2. $G(x) = x^2 + c$ for some $c < -2$ (you choose c).
3. The doubling function D.

For each of these functions, choose ten different initial seeds. For the doubling function, each seed should be in the interval $[0, 1)$. For $F(x) = x^2 - 2$, each should be in the interval $(-2, 2)$. For each chosen seed, compute the first 100 points on the corresponding orbit. Record the results by listing the initial seed together with what happened to the orbit; that is, determine whether the orbit is fixed, periodic, eventually periodic, or has no visible pattern.

Results: After collecting all your data, write a brief essay summarizing what you have seen for each function. Given your data, can you make any conjectures about the behavior of these functions? For a given function, do all (or almost all) orbits behave in the same way? Include in your essay your conjectures and speculations about each function.

Notes and Questions:
1. For the doubling function, you should try certain rational values such as $x_0 = \frac{1}{5}$ and $x_0 = \frac{1}{9}$ whose orbits we already know. Do your results agree with what we already know? Or does the computer give false results? If so, can you explain why this happens? In a further essay, discuss the results of this experiment together with your theory for why the computer reacted the way it did.

2. As a hint, consider what would happen if your computer actually stored numbers in binary form. The computer would not store the entire binary expansion of a given number; rather, it would truncate this expansion so that only an approximation to the number is stored. That is, numbers in the interval $0 \le x_0 < 1$ would be stored in the computer as a finite sum of the form*

$$\frac{a_1}{2} + \frac{a_2}{2^2} + \cdots + \frac{a_n}{2^n}.$$

What happens to such numbers under iteration of the doubling function?

3. If you have access to a program that does exact rational arithmetic, use this to compute orbits of rational numbers of the form p/q under the doubling function. What do you now see? Can you make any conjectures about orbits of rational points?

Exercises

1. Let $F(x) = x^2$. Compute the first five points on the orbit of $1/2$.

2. Let $F(x) = x^2 + 1$. Compute the first five points on the orbit of 0.

3. Let $F(x) = x^2 - 2$. Compute $F^2(x)$ and $F^3(x)$.

4. Let $S(x) = \sin(2x)$. Compute $S^2(x), S^3(x)$, and $S^4(x)$.

5. Let $F(x) = x^2$. Compute $F^2(x), F^3(x)$, and $F^4(x)$. What is the formula for $F^n(x)$?

6. Let $A(x) = |x|$. Compute $A^2(x)$ and $A^3(x)$.

---

*   In practice, the computer would most likely store the number in floating point form, which is a slight variation on this theme.

7. Find all real fixed points (if any) for each of the following functions:

    a. $F(x) = 3x + 2$

    b. $F(x) = x^2 - 2$

    c. $F(x) = x^2 + 1$

    d. $F(x) = x^3 - 3x$

    e. $F(x) = |x|$

    f. $F(x) = x^5$

    g. $F(x) = x^6$

    h. $F(x) = x \sin x$

8. What are the eventually fixed points for $A(x) = |x|$?

9. Let $F(x) = 1 - x^2$. Show that 0 lies on a 2-cycle for this function.

10. Consider the function $F(x) = |x - 2|$. $|x-2|-x$

    a. What are the fixed points for F?

    b. If m is an odd integer, what can you say about the orbit of m?

    c. What happens to the orbit if m is even?

The following four exercises deal with the doubling function D.

11. For each of the following seeds, discuss the behavior of the resulting orbit under D.

    a. $x_0 = 0.3$

    b. $x_0 = 0.7$

    c. $x_0 = 1/8$

    d. $x_0 = 1/16$

    e. $x_0 = 1/7$

    f. $x_0 = 1/14$

    g. $x_0 = 1/11$

    h. $x_0 = 3/22$

12. Give an explicit formula for $D^2(x)$ and $D^3(x)$. Can you write down a general formula for $D^n(x)$?

13. Sketch the graph of D^2 and D^3. What will the graph of D^n look like?

14. Using your answer to exercise 12, find all fixed points for D^2 and D^3. How many fixed points do D^4 and D^5 have? What about D^n?

The following five exercises deal with the function

$$T(x) = \begin{cases} 2x & \text{if } 0 \le x \le 1/2 \\ 2 - 2x & \text{if } 1/2 < x \le 1. \end{cases}$$

The function T is called a *tent map* because of the shape of its graph on the interval $[0, 1]$.

15. Find a formula for $T^2(x)$.

16. Sketch the graphs of T and T^2.

17. Find all fixed points for T and T^2.

18. Find an explicit formula for $T^3(x)$ and sketch the graph of T^3.

19. What does the graph of T^n look like?

CHAPTER 4

Graphical Analysis

In this section we introduce a geometric procedure that will help us understand the dynamics of one-dimensional mappings. This procedure, called *graphical analysis*, enables us to use the graph of a function to determine the behavior of orbits in many cases.

4.1   Graphical Analysis

Suppose we have the graph of a function F and wish to display the orbit of a given point x_0. We begin by superimposing the diagonal line $y = x$ on the graph of F. As we saw in Section 3.3, the points of intersection of the diagonal with the graph give us the fixed points of F. To find the orbit of x_0, we begin at the point (x_0, x_0) on the diagonal directly above x_0 on the x-axis. We first draw a vertical line to the graph of F. When this line meets the graph, we have reached the point $(x_0, F(x_0))$. We then draw a horizontal line from this point to the diagonal. We reach the diagonal at the point whose y-coordinate is $F(x_0)$, and so the x-coordinate is also $F(x_0)$. Thus we reach the diagonal directly over the point whose x-coordinate is $F(x_0)$, the next point on the orbit of x_0.

Now we continue this procedure. Draw a vertical line from $(F(x_0), F(x_0))$ on the diagonal to the graph: this yields the point $(F(x_0), F^2(x_0))$. Then a horizontal line to the diagonal reaches the diagonal at $(F^2(x_0), F^2(x_0))$, directly above the next point in the orbit.

To display the orbit of x_0 geometrically, we thus continue this procedure: we first draw a vertical line from the diagonal to the graph, then a horizontal line from the graph back to the diagonal. The resulting "staircase" or "cobweb" provides an illustrative picture of the orbit of x_0.

Figure 4.1 shows a typical application of graphical analysis. This procedure may be used to describe some of the dynamical behavior we saw in the previous section. For example, in Figure 4.2 we sketch graphical analysis of $F(x) = \sqrt{x}$. Note that any positive x_0 gives a staircase which leads to the point of intersection of the graph of F with the diagonal. This is, of course, the fixed point at $x = 1$.

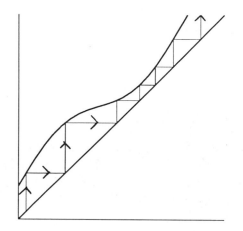

Fig. 4.1 Graphical analysis.

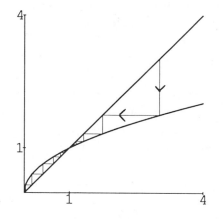

Fig. 4.2 Graphical analysis of $F(x) = \sqrt{x}$.

In Figure 4.3 we depict graphical analysis of $C(x) = \cos x$. Note that any orbit in this case tends again to the point of intersection of the graph of C with the diagonal. As we observed numerically in the previous section, this point is given approximately by $0.73908\ldots$ (in radians).

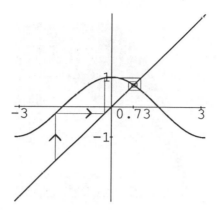

Fig. 4.3 Graphical analysis of $C(x) = \cos x$.

As we saw in the previous section, periodic points for F satisfy $F^n(x_0) = x_0$. This means that the line segments generated by graphical analysis eventually return to (x_0, x_0) on the diagonal, thus yielding a closed "circuit" in the graphical analysis. Figure 4.4a shows that $F(x) = x^2 - 1.1$ admits a 2-cycle as illustrated by the square generated by graphical analysis. Figure 4.4b shows that many orbits tend to this cycle. This 2-cycle can be computed explicitly. See exercise 6.

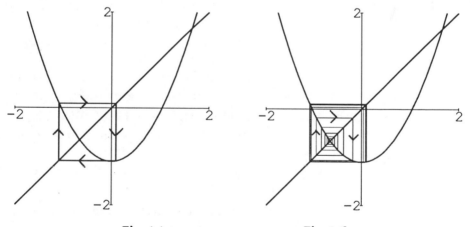

Fig. 4.4a   **Fig. 4.4b**
Graphical analysis of $F(x) = x^2 - 1.1$.

We cannot decipher the behavior of all orbits by means of graphical analysis. For example, in Figure 4.5 we have applied graphical analysis to the quadratic function $F(x) = 4x(1-x)$. Note how complicated the orbit of

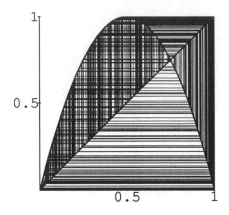

Fig. 4.5 Graphical analysis of $F(x) = 4x(1-x)$.

x_0 is! This is another glimpse of chaotic behavior.

4.2   Orbit Analysis

Graphical analysis sometimes allows us to describe the behavior of *all* orbits of a dynamical system. For example, consider the function $F(x) = x^3$. The graph of F shows that there are three fixed points: at $0, 1$, and -1. These are the solutions of the equation $x^3 = x$, or $x^3 - x = 0$. Graphical analysis then allows us to read off the following behavior. If $|x_0| < 1$ then the orbit of x_0 tends to zero as depicted in Figure 4.6a. On the other hand, if $|x_0| > 1$, then the orbit of x_0 tends to $\pm\infty$, as in Figure 4.6b.

Thus we see that we have accounted for the behavior of the orbits of all points. When we can do this, we say that we have performed a complete *orbit analysis*. While this is not always possible, we will strive in the sequel to develop techniques that will allow us to understand as many of the orbits of a dynamical system as possible.

At this juncture, we should emphasize that graphical analysis is by no means a completely rigorous tool; in most cases we cannot use graphical analysis as a proof that certain dynamical phenomena occur. However, there is no question that this procedure is a valuable tool that helps explain what is going on.

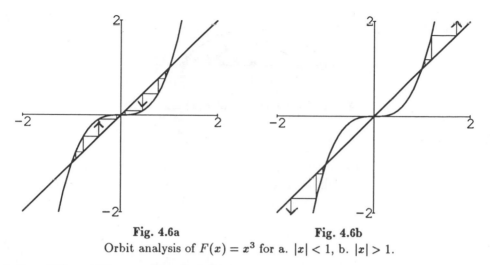

Fig. 4.6a   **Fig. 4.6b**
Orbit analysis of $F(x) = x^3$ for a. $|x| < 1$, b. $|x| > 1$.

4.3   The Phase Portrait

One succinct method for depicting all orbits of a dynamical system is the *phase portrait* of the system. This is a picture on the real line of the orbits. In the one-dimensional case, the phase portrait gives us no more information than graphical analysis. In dimension two, however, when graphical analysis is no longer possible, we will rely solely on the phase portrait to describe the behavior of orbits.

In the phase portrait, we represent fixed points by solid dots and the dynamics along orbits by arrows. For example, as we saw above, for $F(x) = x^3$, the fixed points occur at $0, \pm 1$. If $|x_0| < 1$, then $F^n(x_0) \to 0$, whereas if $|x_0| > 1$, $F^n(x_0) \to \pm \infty$. The phase portrait for this map is shown in Figure 4.7.

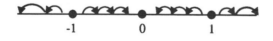

-1   0   1

Fig. 4.7 Phase portrait of $F(x) = x^3$.

As another example, $F(x) = x^2$ has two fixed points, at 0 and 1, and an eventually fixed point at -1. Note that if $x_0 < 0$, then $F(x_0) > 0$ and all subsequent points on the orbit of x_0 are positive. The phase portrait of $F(x) = x^2$ is shown in Figure 4.8.

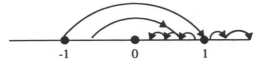

Fig. 4.8 Phase portrait of $F(x) = x^2$.

Exercises

1. Use graphical analysis to describe the fate of all orbits for each of the following functions. Use different colors for orbits that behave differently.

    a. $F(x) = 2x$
    b. $F(x) = \frac{1}{3x}$
    c. $F(x) = -2x + 1$
    d. $F(x) = x^2$
    e. $F(x) = -x^3$
    f. $F(x) = x - x^2$
    g. $S(x) = \sin x$

2. Use graphical analysis to find $\{x_0 \mid F^n(x_0) \to \pm\infty\}$ for each of the following functions.

    a. $F(x) = 2x(1 - x)$
    b. $F(x) = x^2 + 1$
    c. $T(x) = \begin{cases} 2x & x \leq 1/2 \\ 2 - 2x & x > 1/2 \end{cases}$

3. Sketch the phase portraits for each of the functions in exercise 1.

4. Perform a complete orbit analysis for each of the following functions.
    a. $F(x) = \frac{1}{2}x - 2$
    b. $A(x) = |x|$
    c. $F(x) = -x^2$
    d. $F(x) = -x^5$
    e. $F(x) = 1/x$
    f. $E(x) = e^x$

5. Let $F(x) = |x - 2|$. Use graphical analysis to display a variety of orbits of F. Use red to display cycles of period 2, blue for eventually fixed orbits, and green for orbits that are eventually periodic.

6. Consider $F(x) = x^2 - 1.1$. First find the fixed points of F. Then use the fact that these points are also solutions of $F^2(x) = x$ to find the cycle of prime period 2 for F.

7. All of the following exercises deal with dynamics of linear functions of the form $F(x) = ax + b$ where a and b are constants.

   a. Find the fixed points of $F(x) = ax + b$.

   b. For which values of a and b does F have no fixed points?

   c. For which values of a and b does F have infinitely many fixed points?

   d. For which values of a and b does F have exactly one fixed point?

   e. Suppose F has just one fixed point and $0 < |a| < 1$. Using graphical analysis, what can you say about the behavior of all other orbits of F? We will call these fixed points *attracting* fixed points later. Why do we use this terminology?

   f. What is the behavior of all orbits when $a = 0$?

   g. Suppose F has just one fixed point and $|a| > 1$. Using graphical analysis, what can you say about the behavior of all other orbits of F in this case? We call such fixed points *repelling*. Can you explain why?

   h. Perform a complete orbit analysis for $F(x) = x + b$ in case $b > 0$, $b = 0$, and $b < 0$.

   i. Perform a complete orbit analysis for $F(x) = -x + b$.

CHAPTER 5

Fixed and Periodic Points

Fixed points and cycles are among the most important kinds of orbits of a dynamical system, so it is important that we be able to find them easily. As we have seen, sometimes this involves solving equations or drawing accurate graphs. These methods, however, are not always easily carried out. As we shall see in this chapter, calculus comes to the rescue.

5.1   A Fixed Point Theorem

One of the simplest criteria for finding fixed points is an immediate consequence of the following important fact from calculus:

The Intermediate Value Theorem. *Suppose $F: [a, b] \to \mathbf{R}$ is continuous. Suppose y_0 lies between $F(a)$ and $F(b)$. Then there is an x_0 in the interval $[a, b]$ with $F(x_0) = y_0$.*

Simply stated, this theorem tells us that a continuous function assumes all values between $F(a)$ and $F(b)$ on the interval $[a, b]$. An immediate consequence is:

Fixed Point Theorem. *Suppose $F: [a, b] \to [a, b]$ is continuous. Then there is a fixed point for F in $[a, b]$.*

Remarks:

1. This theorem asserts the existence of at least one fixed point for F in $[a, b]$; there may of course be more. For example, *all* points in any interval $[a, b]$ are fixed by the identity function $F(x) = x$.

2. There are several important hypotheses in this theorem, the first two being continuity and the fact that F takes the interval $[a, b]$ into itself. Violation of either of these may yield a function without fixed points.

3. Also, it is important that the interval $[a, b]$ be closed. For example, $F(x) = x^2$ takes the interval $(0, 1/2)$ inside itself and is continuous, but there are no fixed points in this open interval (the fixed point 0 lies outside of $(0, 1/2)$).

4. While the Fixed Point Theorem asserts the existence of at least one fixed point, it unfortunately does not give us any method of actually finding this point. However, in practice, we often don't need to know exactly where the fixed point lies. Just the knowledge that it is present in a certain interval often suffices for our purposes.

The proof of the Fixed Point Theorem follows from the Intermediate Value Theorem applied to $H(x) = F(x) - x$. This is a continuous function that satisfies

$$H(a) = F(a) - a \geq 0$$
$$H(b) = F(b) - b \leq 0.$$

Thus there is a c in the interval $[a, b]$ with $H(c) = 0$. This c satisfies $F(c) - c = 0$ and so is our fixed point.

5.2   Attraction and Repulsion

There are two markedly different types of fixed points, *attracting* and *repelling* fixed points. We will make these notions precise in a moment, but for now the idea behind these concepts is illustrated by the squaring function $F(x) = x^2$. This map has two fixed points, at 0 and at 1. But note what happens to the orbits of nearby points. If we choose any x_0 with $|x_0| < 1$, then the orbit of x_0 rapidly approaches zero. For example, the orbit of 0.1 is

$$0.1, \ 0.01, \ 0.0001, \ 0.00000001, \ldots.$$

In fact, any x_0 with $0 \leq x_0 < 1$, no matter how close to 1, leads to an orbit that tends "far" from 1 and close to 0. For example, the orbit of 0.9 is

$$0.9, 0.81, 0.6561, 0.430467\ldots, 0.185302\ldots, 0.034336\ldots, 0.00117\ldots.$$

More precisely, if $0 \leq x_0 < 1$, then $F^n(x_0) \to 0$ as $n \to \infty$. On the other hand, if $x_0 > 1$, then again the orbit moves far from 1. For example, the orbit of 1.1 is

$$1.1, 1.21, 1.4641, 2.1436\ldots, 4.5950\ldots, 21.114\ldots, 445.79\ldots.$$

Thus, if $x_0 > 1$, we have $F^n(x_0) \to \infty$ as $n \to \infty$ and hence the orbit tends far from 1.

Clearly, points that are close to 0 have orbits that are attracted to 0, while points close to 1 have orbits that are repelled, that move far from 1. Graphical analysis of this function, as in Figure 5.1, shows vividly the difference between these two types of fixed points.

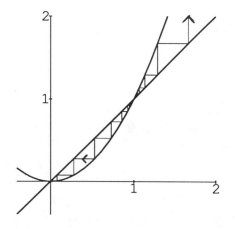

Fig. 5.1 The point 0 is an attracting fixed point for $F(x) = x^2$, while 1 is a repelling fixed point.

5.3   Calculus of Fixed Points

As another example, consider the linear functions $A(x) = \alpha x$ with $0 < \alpha < 1$ and $B(x) = \beta x$ with $\beta > 1$. Each function has a fixed point at 0, but this fixed point is attracting for A and repelling for B. Graphical analysis shows this clearly in Figure 5.2.

Now consider the same functions, this time with $-1 < \alpha < 0$ and $\beta < -1$. Again, graphical analysis (Figure 5.3) shows that A has an attracting fixed point at 0 and B has a repelling fixed point at 0. This time orbits hop back and forth from left to right about zero, but still they either approach zero or are repelled from zero. Clearly, the slope of these straight lines plays a crucial role in determining whether or not a linear function has an attracting or repelling fixed point. As you saw in the exercises in the last chapter, this fact holds for functions of the form $cx + d$ where c and d are constants as well.

This suggests that calculus will allow us to differentiate (pardon the lousy pun) between attracting and repelling fixed points for nonlinear functions,

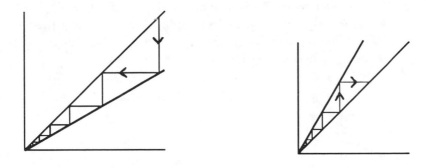

(a)   (b)

Fig. 5.2 Graphical analysis of **(a)** $A(x) = \alpha x$, $0 < \alpha < 1$ and **(b)** $B(x) = \beta x$, $\beta > 1$.

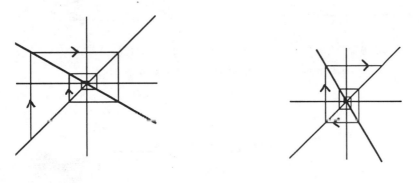

(a)   (b)

Fig. 5.3 Graphical analysis of **(a)** $A(x) = \alpha x$, $-1 < \alpha < 0$ and **(b)** $B(x) = \beta x$, $\beta < -1$.

for the derivative gives us the slope of the tangent line to the graph of a function. In particular, near a fixed point x_0, if we examine the graph of the function closely, the magnitude of the first derivative $F'(x_0)$ tells us how steeply the graph crosses the diagonal at this point. This leads to an important definition:

Definition. Suppose x_0 is a fixed point for F. Then x_0 is an *attracting fixed point* if $|F'(x_0)| < 1$. The point x_0 is a *repelling fixed point* if $|F'(x_0)| > 1$. Finally, if $|F'(x_0)| = 1$, the fixed point is called *neutral* or *indifferent*.

The geometric rationale for this terminology is supplied by graphical analysis. Consider the graphs in Figure 5.4a,b. Both of these functions have fixed points at x_0. The slope of the tangent line at x_0, $F'(x_0)$, is in both cases less than 1 in magnitude: $|F'(x_0)| < 1$. Note that this forces nearby orbits to approach x_0, just as in the linear cases above. If $-1 < F'(x_0) < 0$, as in Figure 5.4b, the orbit hops from one side of x_0 to the other as it approaches x_0. The phase portraits in the two cases $0 < F'(x_0) < 1$ and $-1 < F'(x_0) < 0$ are sketched in Figure 5.5.

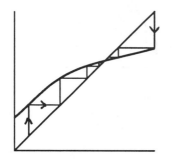 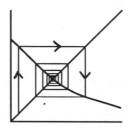

(a) $0 < F'(x_0) < 1$ **(b)** $-1 < F'(x_0) < 0$

Fig. 5.4 In each case, x_0 is an attracting fixed point

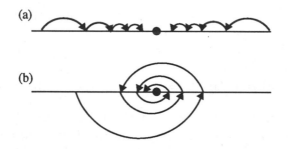

(a) $0 < F'(x_0) < 1$ **(b)** $-1 < F'(x_0) < 0$

Fig. 5.5 Possible phase portraits near an attracting fixed point.

On the other hand, if $|F'(x_0)| > 1$, graphical analysis shows that nearby points have orbits that move farther away, that is, are repelled. Again, if $F'(x_0) < -1$, orbits oscillate from side to side of x_0 as they move away, as in Figure 5.6. As before, the phase portraits (Figure 5.7) show how nearby orbits are repelled in these cases.

As an example, consider the function $F(x) = 2x(1 - x) = 2x - 2x^2$. Clearly, 0 and 1/2 are fixed points for F. We have $F'(x) = 2 - 4x$, so $F'(0) = 2$ and $F'(1/2) = 0$. Thus 0 is a repelling fixed point, while 1/2 is attracting. Graphical analysis confirms this, as shown in Figure 5.8.

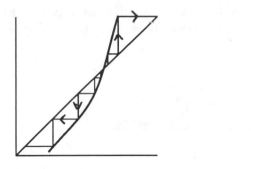
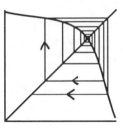

(a) $F'(x_0) > 1$ **(b)** $F'(x_0) < -1$

Fig. 5.6 In each case, x_0 is a repelling fixed point.

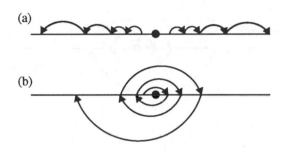

Fig. 5.7 Phase portraits near a repelling fixed point.

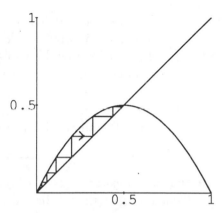

Fig. 5.8 $F(x) = 2x(1 - x)$ has an attracting fixed point at $1/2$ and a repelling fixed point at 0.

Note that if $F'(x_0) = 0$, we may have several different types of phase portraits, as shown in Figure 5.9. In all cases, however, the fixed point is attracting.

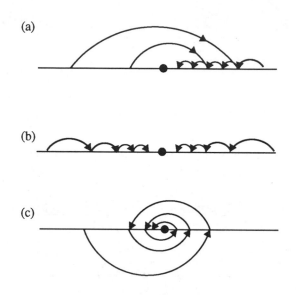

Fig. 5.9 Phase portraits near zero for **(a)** $F(x) = x^2$, **(b)** $F(x) = x^3$, and **(c)** $F(x) = -x^3$. In each case, $F(0) = 0$ and $F'(0) = 0$.

5.4   Why is this true?

The graphical analyses in Figures 5.4 and 5.6 are certainly convincing evidence for the truth of our claim that the magnitude of $F'(x_0)$ determines whether x_0 is attracting or repelling. But how do we know that this is true, especially when the slopes of the tangent lines are close to the cut-off points, namely ± 1?

The answer is provided by one of the most useful theorems in all of calculus, the Mean Value Theorem. Recall the fairly technical statement of this result:

The Mean Value Theorem. *Suppose F is a differentiable function on the interval $a \leq x \leq b$. Then there exists c between a and b for which the following equation is true:*

$$F'(c) = \frac{F(b) - F(a)}{b - a}.$$

The content of this theorem is best exhibited geometrically. The quantity

$$M = \frac{F(b) - F(a)}{b - a}$$

is the slope of the straight line connecting the two points $(a, F(a))$ and $(b, F(b))$ on the graph of F. So the theorem simply says that, provided F is differentiable on the interval $a \leq x \leq b$, there is some point c between a and b at which the slope of the tangent line, $F'(c)$, is exactly equal to M. This is displayed geometrically in Figure 5.10.

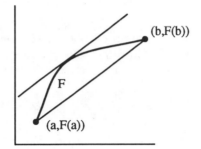

Fig. 5.10 The Mean Value Theorem.
The slope of the tangent line at c, $F'(c)$, is equal to M.

For our purposes, the importance of the Mean Value Theorem lies in its two corollaries:

Attracting Fixed Point Theorem. *Suppose x_0 is an attracting fixed point for F. Then there is an interval I that contains x_0 in its interior and in which the following condition is satisfied: if $x \in I$, then $F^n(x) \in I$ for all n and, moreover, $F^n(x) \to x_0$ as $n \to \infty$.*

Proof: Since $|F'(x_0)| < 1$, there is a number $\lambda > 0$ such that $|F'(x_0)| < \lambda < 1$. We may therefore choose a number $\delta > 0$ so that $|F'(x)| < \lambda$ provided x belongs to the interval $I = [x_0 - \delta, x_0 + \delta]$. Now let p be any point in I. By the Mean Value Theorem

$$\frac{|F(p) - F(x_0)|}{|p - x_0|} < \lambda,$$

so that

$$|F(p) - F(x_0)| < \lambda|p - x_0|.$$

Since x_0 is a fixed point, it follows that

$$|F(p) - x_0| < \lambda|p - x_0|.$$

This means that the distance from $F(p)$ to x_0 is smaller than the distance from p to x_0, since $0 < \lambda < 1$. In particular, $F(p)$ also lies in the interval I. Therefore we may apply the same argument to $F(p)$ and $F(x_0)$, finding

$$
\begin{aligned}
|F^2(p) - x_0| &= |F^2(p) - F^2(x_0)| \\
&< \lambda|F(p) - F(x_0)| \\
&< \lambda^2|p - x_0|.
\end{aligned}
$$

Since $\lambda < 1$, we have $\lambda^2 < \lambda$. This means that the points $F^2(p)$ and x_0 are even closer together than $F(p)$ and x_0. Thus we may continue using this argument to find that, for any $n > 0$,

$$|F^n(p) - x_0| < \lambda^n|x - x_0|.$$

Now $\lambda^n \to 0$ as $n \to \infty$. Thus, $F^n(p) \to x_0$ as $n \to \infty$. This completes the proof.

We observe that the proof of the Attracting Fixed Point Theorem actually tells us a little bit more than the statement of the theorem implies. The inequality

$$|F^n(p) - x_0| < \lambda^n|x - x_0|$$

says that nearby orbits (those beginning in I) actually converge *exponentially* to the fixed point. Notice that the number λ may be chosen very close to $|F'(x_0)|$, so it is essentially this quantity that governs the rate of approach of orbits to x_0.

Arguing in an entirely analogous manner, we have

Repelling Fixed Point Theorem. *Suppose x_0 is a repelling fixed point for F. Then there is an interval I that contains x_0 in its interior and in which the following condition is satisfied: if $x \in I$ and $x \neq x_0$, then there is an integer $n > 0$ such that $F^n(x) \notin I$.*

These two theorems combined justify our use of the terminology "attracting" and "repelling" to describe the corresponding fixed points. In particular, they tell us the "local" dynamics near any fixed point x_0 for which $|F'(x_0)| \neq 1$.

One major difference between attracting and repelling fixed points is the fact that attracting points are "visible" on the computer, whereas repelling fixed points generally are not. We can often find an attracting fixed point by choosing an initial seed randomly and computing its orbit numerically. If this orbit ever enters the interval I about an attracting fixed point, then we know the fate of this orbit—it necessarily converges to the attracting fixed point. On the other hand, in the case of a repelling fixed point, the randomly chosen orbit would have to land exactly on the fixed point in order for us to see it. This rarely happens, for even if the orbit comes very close to a repelling fixed point, roundoff error will throw us off this fixed point and onto an orbit that moves away.

The situation for a neutral fixed point is not nearly as simple as the attracting or repelling cases. For example, the identity function $F(x) = x$ fixes all points, but none are attracting or repelling. Also, $F(x) = -x$ fixes zero, but this is not an attracting or repelling fixed point since all other points lie on cycles of period 2. Finally, as Figure 5.11 shows, $F(x) = x - x^2$ has a fixed point at zero which is attracting from the right but repelling from the left. Note that $|F'(0)| = 1$ in all three cases.

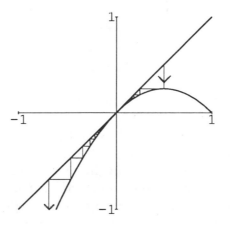

Fig. 5.11 The fixed point at zero for $F(x) = x - x^2$ is neither attracting nor repelling.

On the other hand, neutral fixed points may attract or repel *all* nearby orbits. For example, graphical analysis shows that $F(x) = x - x^3$ has a fixed point that attracts the orbit of any x with $|x| < 1$, whereas $F(x) = x + x^3$ repels all orbits away from 0. These fixed points are sometimes called *weakly* attracting or *weakly* repelling, since the convergence or divergence is quite slow. This is illustrated in Experiment 5.6.

5.5   Periodic Points

Just as in the case of fixed points, periodic points may also be classified as attracting, repelling, or neutral. The calculus here is only slightly more complicated.

Let's begin with an example. The function $F(x) = x^2 - 1$ has an attracting cycle of period 2 with orbit $0, -1, 0, -1, \ldots$. Graphical analysis as depicted in Figure 5.12 indicates that this cycle should be attracting.

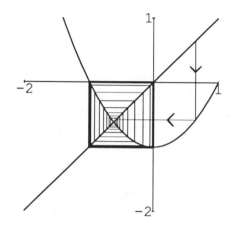

Fig. 5.12 An attracting cycle of period 2 for $F(x) = x^2 - 1$.

To see why this is the case, we examine the graph of $F^2(x) = (x^2 - 1)^2 - 1 = x^4 - 2x^2$, as shown in Figure 5.13. Note that F^2 has four fixed points: at the two fixed points of F as well as at the period-2 points 0 and -1. Note that the derivative of F^2, namely $4x^3 - 4x$, vanishes at both 0 and -1, that is, $(F^2)'(0) = (F^2)'(-1) = 0$. This indicates that these two points are attracting fixed points for the second iterate of F. That is, under iteration of F^2, orbits of points nearby 0 or -1 converge to these points. Under iteration of F, however, these orbits cycle back and forth as they converge to the 2-cycle.

These ideas motivate us to extend the definitions of attraction and repulsion to cycles in the natural way: a periodic point of period n is attracting (repelling) if it is an attracting (repelling) fixed point for F^n. This immediately brings up the question of whether periodic orbits can contain some points that are attracting and some that are repelling. As we will see below, calculus says that this is not the case.

To determine if a periodic point x_0 of period n is attracting or repelling, we must compute the derivative of F^n at x_0. Recalling that F^n is the nth

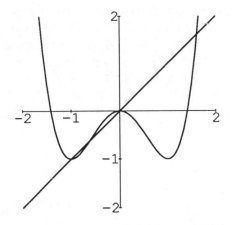

Fig. 5.13 The graph of the second iterate of $F(x) = x^2 - 1$.

iterate of F, not the nth power of F, we see that this demands the use of the Chain Rule. From calculus, we know that if F and G are differentiable functions, then the derivative of their composition is given by

$$(F \circ G)'(x) = F'(G(x)) \cdot G'(x).$$

In particular,

$$(F^2)'(x_0) = F'(F(x_0)) \cdot F'(x_0)$$
$$= F'(x_1) \cdot F'(x_0),$$

and

$$(F^3)'(x_0) = F'(F^2(x_0)) \cdot (F^2)'(x_0)$$
$$= F'(x_2) \cdot F'(x_1) \cdot F'(x_0).$$

Invoking the Chain Rule $n - 1$ times, we find:

Chain Rule Along a Cycle. *Suppose $x_0, x_1, \ldots, x_{n-1}$ lie on a cycle of period n for F with $x_i = F^i(x_0)$. Then*

$$(F^n)'(x_0) = F'(x_{n-1}) \cdot \ldots \cdot F'(x_1) \cdot F'(x_0).$$

Note that this formula tells us that the derivative of F^n at x_0 is simply the product of the derivatives of F at all points on the orbit. This means that we do not have to first compute the formula for F^n. All we need find is the derivative of F and then plug in $x_0, x_1, \ldots, x_{n-1}$ respectively and multiply all of these numbers together. This saves a lot of algebra, especially if n is large. For example, we saw above that for the function $F(x) = x^2 - 1$ that has a 2-cycle at 0 and -1, $(F^2)'(0) = 0$. We computed this by first finding the formula for F^2 and then differentiating at 0. But, since $F'(x) = 2x$, we have quite simply $F'(0) = 0$ and $F'(-1) = -2$, and so, by the Chain Rule Along a Cycle, again $(F^2)'(0) = -2 \cdot 0 = 0$. Similarly, $(F^2)'(-1) = -2 \cdot 0 = 0$. The fact that, in this example, $(F^2)'(0) = (F^2)'(-1)$ is no accident because of the following:

Corollary. *Suppose $x_0, x_1, \ldots x_{n-1}$ lie on an n-cycle for F. Then*

$$(F^n)'(x_0) = (F^n)'(x_1) = \cdots = (F^n)'(x_{n-1}).$$

This corollary follows immediately from the Chain Rule Along a Cycle, since the points on the cycle are exactly the same, no matter which point is chosen as the initial seed.

Example. Consider $F(x) = -\frac{3}{2}x^2 + \frac{5}{2}x + 1$. The point 0 lies on a cycle of period 3 since $F(0) = 1$, $F(1) = 2$, and $F(2) = 0$. We have $F'(x) = -3x + \frac{5}{2}$, so $F'(0) = \frac{5}{2}$, $F'(1) = -\frac{1}{2}$, and $F'(2) = -\frac{7}{2}$. Hence

$$\left(F^3\right)'(0) = F'(2) \cdot F'(1) \cdot F'(0)$$

$$= \left(\frac{-7}{2}\right)\left(\frac{-1}{2}\right)\left(\frac{5}{2}\right)$$

$$= \frac{35}{8} > 1.$$

Therefore this cycle is repelling.

5.6 Experiment: Rates of Convergence

Goal: In this experiment you will investigate how quickly (or slowly) an orbit is attracted to a fixed point. Your goal is to relate this speed of convergence to the value of the derivative at the fixed point.

Procedure: Each of the functions listed below has a fixed point and the orbit of 0.2 is attracted to this point. For each function, use the computer to compute the orbit of 0.2 until it "reaches" the fixed point, or else comes within some designated distance from the fixed point, say 10^{-5}. This exact number will be provided by your instructor since it depends on the type of equipment and software you are using to compute orbits. We say "reaches" in quotes because, in actuality, the orbit will never reach the fixed point. However, because of computer round-off, it may appear that the orbit has landed on the fixed point.

For each function, you should record:
a. The fixed point p.
b. $|F'(p)|$
c. Is p attracting or neutral?
d. The number of iterations necessary for the orbit of 0.2 to "reach" p.

The functions you should try are:

a. $F(x) = x^2 + 0.25$
b. $F(x) = x^2$
c. $F(x) = x^2 - 0.24$
d. $F(x) = x^2 - 0.75$
e. $F(x) = 0.4x(1 - x)$
f. $F(x) = x(1 - x)$
g. $F(x) = 1.6x(1 - x)$
h. $F(x) = 2x(1 - x)$
i. $F(x) = 2.4x(1 - x)$
j. $F(x) = 3x(1 - x)$
k. $F(x) = 0.4 \sin x$
l. $F(x) = \sin x$

Results: After compiling the data, compare the results for each function. In an essay, describe what you have observed. What is the relationship between $|F'(p)|$ and the speed of convergence to p? Which fixed points have fastest convergence and which have slowest?

Notes and Questions: Each of the following functions has a cycle of period 2 that attracts the orbit of 0.2. In each case, determine the cycle experimentally and estimate the derivative of F^2 along the cycle. (You need not solve the equations explicitly to get exact values, although this can be done.) Again compare the attracting vs. neutral cases.

a. $F(x) = x^2 - 1$
b. $F(x) = x^2 - 1.1$
c. $F(x) = x^2 - 1.25$. *Hint:* The 2-cycle is given by $(-1 \pm \sqrt{2})/2$.

Exercises

1. For each of the following functions, find all fixed points and classify them as attracting, repelling, or neutral.

    a. $F(x) = x^2 - x/2$
    b. $F(x) = x(1 - x)$
    c. $F(x) = 3x(1 - x)$
    d. $F(x) = (2 - x)/10$
    e. $F(x) = x^4 - 4x^2 + 2$
    f. $S(x) = \frac{\pi}{2} \sin x$
    g. $S(x) = -\sin x$
    h. $F(x) = x^3 - 3x$
    i. $A(x) = \arctan x$
    j. $T(x) = \begin{cases} 2x & \text{if } x \leq \frac{1}{2} \\ 2 - 2x & \text{if } x > \frac{1}{2} \end{cases}$
    k. $F(x) = 1/x^2$

2. For each of the following functions, zero lies on a periodic orbit. Classify this orbit as attracting, repelling, or neutral.

    a. $F(x) = 1 - x^2$
    b. $C(x) = \frac{\pi}{2} \cos x$
    c. $F(x) = -\frac{1}{2}x^3 - \frac{3}{2}x^2 + 1$
    d. $F(x) = |x - 2| - 1$
    e. $A(x) = -\frac{4}{\pi} \arctan(x + 1)$
    f. $F(x) = \begin{cases} x + 1 & \text{if } x \leq 3.5 \\ 2x - 8 & \text{if } x > 3.5 \end{cases}$

3. Suppose x_0 lies on a cycle of prime period n for the doubling function D. Evaluate $(D^n)'(x_0)$. Is this cycle attracting or repelling?

4. Each of the following functions has a neutral fixed point. Find this fixed point and, using graphical analysis with an accurate graph, determine if it is weakly attracting, weakly repelling, or neither.

    a. $F(x) = x + x^2$
    b. $F(x) = 1/x$
    c. $E(x) = e^{x-1}$ (Fixed point is $x_0 = 1$)
    d. $S(x) = \sin x$
    e. $T(x) = \tan x$
    f. $F(x) = x + x^3$
    g. $F(x) = x - x^3$

   h. $F(x) = -x + x^3$
   i. $F(x) = -x - x^3$
   j. $E(x) = -ee^x$ (fixed point is $x_0 = -1$). *Hint:* Examine in detail the graph of $E^2(x)$ near $x = -1$ using higher derivatives of E^2.
   k. $L(x) = \ln|x - 1|$.

5. Suppose that F has a neutral fixed point at x_0 with $F'(x_0) = 1$. Suppose also that $F''(x_0) > 0$. What can you say about x_0: is x_0 weakly attracting, weakly repelling, or neither? Use graphical analysis and the concavity of the graph of F near x_0 to support your answer.

6. Repeat exercise 5, but this time assume that $F''(x_0) < 0$.

7. Suppose that F has a neutral fixed point at x_0 with $F'(x_0) = 1$ and $F''(x_0) = 0$. Suppose also that $F'''(x_0) > 0$. Use graphical analysis and the concavity of the graph of F near x_0 to show that x_0 is weakly repelling.

8. Repeat exercise 7, but this time assume that $F'''(x_0) < 0$. Show that x_0 is weakly attracting.

9. Combine the results of exercises 5–8 to state a *Neutral Fixed Point Theorem*.

CHAPTER 6

Bifurcations

In this chapter we begin the study of the quadratic family of functions $Q_c(x) = x^2 + c$ where c is a constant. While these functions look simple enough, we will see that their dynamics are amazingly complicated. Indeed, their behavior is not yet completely understood for certain c-values. This chapter is just the beginning of a long story about $Q_c(x) = x^2 + c$ and related dynamical systems—one that will occupy our attention for most of the rest of this book. Here we introduce two of the most important types of bifurcations that occur in dynamics—we will see other types later.

6.1   Dynamics of the Quadratic Map

Throughout this chapter we let Q_c denote the family of quadratic functions $Q_c(x) = x^2 + c$. The number c is a *parameter*—for each different c we get a different dynamical system Q_c. Our goal is to understand how the dynamics of Q_c change as c varies.

As usual, our first task is to find the fixed points of Q_c. These are obtained by solving the quadratic equation

$$x^2 + c = x,$$

which yields two roots:

$$p_+ = \frac{1}{2}\left(1 + \sqrt{1 - 4c}\right)$$

$$p_- = \frac{1}{2}\left(1 - \sqrt{1 - 4c}\right).$$

These are the two fixed points for Q_c.*

Note that p_+ and p_- are real if and only if $1 - 4c \geq 0$, or $c \leq 1/4$. That is, when $c > 1/4$, Q_c has no fixed points whatsoever. When $c = 1/4$, we have $1 - 4c = 0$, so that $p_+ = p_- = 1/2$. Finally, when $c < 1/4$, $1 - 4c > 0$, so that p_+ and p_- are real and distinct. Note that we always have $p_+ > p_-$ in this case.

Let's return now to the case $c > 1/4$. The graph of Q_c is a parabola opening upward. When $c > 1/4$, this graph does not meet the diagonal line $y = x$. Hence graphical analysis shows that all orbits of Q_c, when $c > 1/4$, tend to infinity. Therefore we understand completely the (rather simple) dynamics in this case.

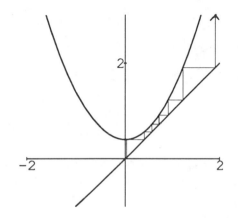

Fig. 6.1 All orbits of $Q_c(x) = x^2 + c$ for $c > 1/4$ tend to infinity.

When c decreases below $1/4$, this situation changes—we encounter our first *bifurcation*. Bifurcation means a division in two, a splitting apart, and that is exactly what has happened to the fixed points of Q_c. For $c > 1/4$, Q_c has no fixed points; for $c = 1/4$, Q_c has exactly one fixed point; but for $c < 1/4$, this fixed point splits into two, one at p_+ and one at p_-. In the next section we will call this kind of bifurcation the *saddle-node* or *tangent bifurcation* and describe it more fully.

Using the results of the previous chapter, we can check whether the fixed points $p_\pm$ are attracting, repelling, or neutral. Since $Q_c'(x) = 2x$, we find

$$Q_c'(p_+) = 1 + \sqrt{1 - 4c}$$

---

*   We really should write $p_+(c)$ and $p_-(c)$ since both of these fixed points depend on c. However, we will usually suppress this added notation for readability.

$$Q_c'(p_-) = 1 - \sqrt{1 - 4c}.$$

Note that $Q_c'(p_+) = 1$ if $c = 1/4$, but $Q_c'(p_+) > 1$ for $c < 1/4$ since $\sqrt{1 - 4c} > 0$ for these c-values. Hence p_+ is a neutral fixed point when $c = 1/4$ but is repelling when $c < 1/4$.

The situation for p_- is slightly more complicated. We have $Q_c'(p_-) = 1$ when $c = 1/4$ (here, of course, $p_- = p_+ = 1/2$). When c is slightly below $1/4$, $Q_c'(p_-) < 1$ so p_- becomes attracting. To find all of the c-values for which $|Q_c'(p_-)| < 1$, we must solve

$$-1 < Q_c'(p_-) < 1$$

or

$$-1 < 1 - \sqrt{1 - 4c} < 1.$$

Solving these inequalities yields

$$2 > \sqrt{1 - 4c} > 0$$

$$4 > 1 - 4c > 0$$

$$-3/4 < c < 1/4.$$

Thus, p_- is an attracting fixed point for Q_c when $-3/4 < c < 1/4$. It is easy to check that, when $c = -3/4$, $Q_c'(p_-) = 1 - \sqrt{1 + 3} = -1$, so p_- is neutral. When $c < -3/4$, $Q_c'(p_-) < -1$, so p_- is repelling.

Let's summarize all of this in a Proposition.

Proposition: The First Bifurcation. *For the family $Q_c(x) = x^2 + c$:*

1. *All orbits tend to infinity if $c > 1/4$.*
2. *When $c = 1/4$, Q_c has a single fixed point at $p_+ = p_- = 1/2$ that is neutral.*
3. *For $c < 1/4$, Q_c has two fixed points at p_+ and p_-. The fixed point p_+ is always repelling.*
   a. *If $-3/4 < c < 1/4$, p_- is attracting.*
   b. *If $c = -3/4$, p_- is neutral.*
   c. *If $c < -3/4$, p_- is repelling.*

We observe one point here that will be useful in the next few sections. For any $c \leq 1/4$, all of the interesting dynamics occurs in the interval $-p_+ \leq x \leq p_+$. Note that $Q_c(-p_+) = p_+$, so $-p_+$ is an eventually fixed point.

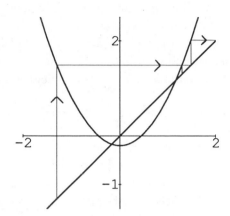

Fig. 6.2 If $c \leq 1/4$, any x with $x > p_+$ or $x < -p_+$
has an orbit that tends to infinity.

Indeed, graphical analysis shows that if $x > p_+$ or $x < -p_+$, then the orbit
of x simply tends to infinity (Figure 6.2).

One can also prove that, for $-3/4 < c < 1/4$, all orbits in the interval
$(-p_+, p_+)$ tend to the attracting fixed point at p_-. Proving this is straightforward when $0 \leq c < 1/4$, but the proof is more complicated in the other case.
We content ourselves by simply illustrating this fact via graphical analysis
in Figure 6.3.

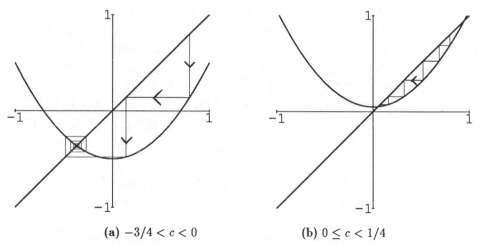

(a) $-3/4 < c < 0$ (b) $0 \leq c < 1/4$

Fig. 6.3 Graphical analysis shows that all orbits of Q_c
in the interval $-p_+ < x < p_+$ tend to p_- when $-3/4 < c < 1/4$.

We now turn our attention to what happens as c decreases below $-3/4$.

From the above we know that the fixed point p_- ceases to be attracting and becomes repelling as this occurs. We also know that there are no other cycles when $c > -3/4$. We ask if this is true for $c < -3/4$? The answer is no: a cycle of period 2 appears when $c < -3/4$. To see this we solve the equation $Q_c^2(x) = x$. The resulting equation is of fourth degree and looks formidable.

$$x^4 + 2cx^2 - x + c^2 + c = 0.$$

However, we know two solutions of this equation, namely the fixed points p_+ and p_-. Hence $x - p_+$ and $x - p_-$ are factors of this equation (for $c < 1/4$). Therefore, so is $(x - p_+)(x - p_-) = x^2 + c - x$. We know what this product is since both p_- and p_+ are fixed points and thus solutions of $x^2 + c - x = 0$.

Thus we may divide by this quantity, obtaining

$$\frac{x^4 + 2cx^2 - x + c^2 + c}{x^2 + c - x} = x^2 + x + c + 1.$$

Then the solutions of $x^2 + x + c + 1 = 0$ give fixed points of Q_c^2 that are also periodic points of period 2 for Q_c. These roots are

$$q_\pm = \frac{1}{2}\left(-1 \pm \sqrt{-4c - 3}\right).$$

Note that $q_\pm$ also depend on c. Furthermore, $q_\pm$ are real if only if $-4c-3 \geq 0$, that is, if $c \leq -3/4$. Hence we have a new kind of bifurcation called a *period-doubling bifurcation*. As c decreases below $-3/4$, two things occur: the fixed point p_- changes from attracting to repelling and a new 2-cycle appears at $q_\pm$. Note that when $c = -3/4$, we have $q_+ = q_- = -1/2 = p_-$, so these two new periodic points originated at p_- when $c = -3/4$.

We will leave the following two observations as exercises. For $-5/4 < c < -3/4$, the points $q_\pm$ lie on an attracting 2-cycle, but for $c < -5/4$, they lie on a repelling 2-cycle. We may therefore summarize this situation as follows.

Proposition: The Second Bifurcation. *For the family $Q_c(x) = x^2 + c$:*
1. *For $-3/4 < c < 1/4$, Q_c has an attracting fixed point at p_- and no 2-cycles.*
2. *For $c = -3/4$, Q_c has a neutral fixed point at $p_- = q_\pm$ and no 2-cycles.*
3. *For $-5/4 < c < -3/4$, Q_c has repelling fixed points at $p_\pm$ and an attracting 2-cycle at $q_\pm$.*

6.2   The Saddle-Node Bifurcation

In the next two sections we temporarily suspend our investigation of the quadratic family to describe more completely the bifurcations encountered in the last section. We will consider a *one-parameter family of functions* F_λ. Here λ is a parameter, so that for each λ, F_λ is a function of x. We will assume that F_λ depends smoothly on λ and x in such a family, unless otherwise noted. For example, $F_\lambda(x) = \lambda x(1 - x)$, $S_\lambda(x) = \lambda \sin(x)$, and $E_\lambda(x) = \exp(x) + \lambda$ are all one-parameter families of functions.

Bifurcations occur in a one-parameter family of functions when there is a change in the fixed or periodic point structure as λ passes through some particular parameter value. Among the most important bifurcations is the saddle-node or tangent bifurcation.

Definition. A one-parameter family of functions F_λ undergoes a *saddle-node (or tangent) bifurcation* at the parameter value λ_0 if there is an open interval I and an $\epsilon > 0$ such that:

1. For $\lambda_0 - \epsilon < \lambda < \lambda_0$, F_λ has no fixed points in the interval I.
2. For $\lambda = \lambda_0$, F_λ has one fixed point in I and this fixed point is neutral.
3. For $\lambda_0 < \lambda < \lambda_0 + \epsilon$, F_λ has two fixed points in I, one attracting and one repelling.

This is a complicated definition. Intuitively, this definition means the following. A saddle-node or tangent bifurcation occurs if the functions F_λ have no fixed points in an interval I for λ-values slightly less than λ_0, exactly one fixed point in I when $\lambda = \lambda_0$, and exactly two fixed points in I for λ slightly larger than λ_0.

Remarks:

1. There is nothing sacred about the inequalities in λ: a saddle-node bifurcation also occurs if the direction of the bifurcation is reversed, that is, no fixed points for $\lambda_0 + \epsilon > \lambda > \lambda_0$ and so forth.

2. Periodic points may undergo a saddle-node bifurcation. These are described by simply replacing F_λ with F_λ^n for a cycle of period n in the above definition.

3. The saddle-node bifurcation typically occurs when the graph of F_{λ_0} has a quadratic tangency with the diagonal at (x_0, x_0) (so $F'_{\lambda_0}(x_0) = 1$ but

$F''_{\lambda_0}(x_0) \neq 0)$. This condition implies that the graph of F_{λ_0} is either concave up or down, so that near x_0, F_{λ_0} has only the one fixed point x_0.*

4. The fact that F_{λ_0} is tangent to the diagonal at x_0 is the reason for the terminology "tangent" bifurcation. The term "saddle-node" comes from a description of this bifurcation in higher dimensions and in the field of differential equations. In our simple setting this terminology is not very transparent, but it is nevertheless standard.

5. Bifurcation theory is a "local" theory in that we are only concerned about changes in the periodic point structure near the parameter value λ_0. That is the reason for the ϵ in the definition. Usually, ϵ is small.

The typical saddle-node bifurcation occurs as depicted in Figure 6.4. The accompanying phase portraits are shown in Figure 6.5.

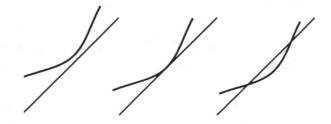

(a) $\lambda < \lambda_0$ (b) $\lambda = \lambda_0$ (c) $\lambda > \lambda_0$

Fig. 6.4 A typical saddle-node bifurcation.

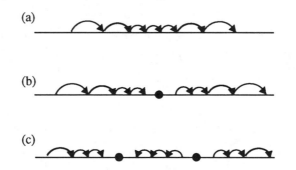

Fig. 6.5 Phase portraits for (a) $\lambda < \lambda_0$, (b) $\lambda = \lambda_0$, and (c) $\lambda > \lambda_0$.

---

*   The precise conditions that guarantee the occurrence of a saddle-node bifurcation may be found in Devaney, R. L. *An Introduction to Chaotic Dynamical Systems, Second Edition.* Addison-Wesley, (1989), page 88.

Example. The quadratic family $Q_c(x) = x^2 + c$ has a saddle-node bifurcation at $c = 1/4$, as discussed in the previous section. In this example we may take the interval I to be the entire real line. There are no fixed points when $c > 1/4$, a neutral fixed point when $c = 1/4$, and a pair of fixed points (one attracting, one repelling) for $-3/4 < c < 1/4$. Thus we may take $\epsilon = 1$ in this example.

Example. The family $E_\lambda(x) = e^x + \lambda$ has a saddle-node bifurcation when $\lambda = -1$. Note that $E_{-1}(0) = 0$, $E'_{-1}(0) = 1$, and $E''_{-1}(0) = 1$. The graphs of E_λ for $\lambda < -1$ and $\lambda > -1$ show how this bifurcation unfolds (Figure 6.6).

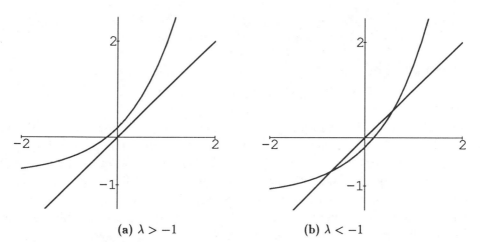

(a) $\lambda > -1$ (b) $\lambda < -1$

Fig. 6.6 A saddle-node bifurcation in the exponential family $E_\lambda(x) = e^x + \lambda$.

Example. The family $F_\lambda(x) = \lambda x(1 - x)$ undergoes a bifurcation when $\lambda = 1$ and $x = 0$, but this is not a saddle-node bifurcation. We do have $F_1(0) = 0$ and $F'_1(0) = 1$, and there is a unique neutral fixed point at $x = 0$. In fact, 0 is a fixed point for any λ. However, for any nonzero λ close to $\lambda = 1$, there is a second fixed point close to 0, as shown in the graphs in Figure 6.7. So this is technically not a saddle-node bifurcation.

To understand bifurcation behavior, it is often helpful to look at the *bifurcation diagram*. This is a picture in the λ, x-plane of the relevant fixed and periodic points as functions of λ. For example, in the quadratic family $Q_c(x) = x^2 + c$, we observed a saddle-node bifurcation at $c = 1/4$. For $c > 1/4$ there were no fixed points, but for $c < 1/4$ we found two fixed points

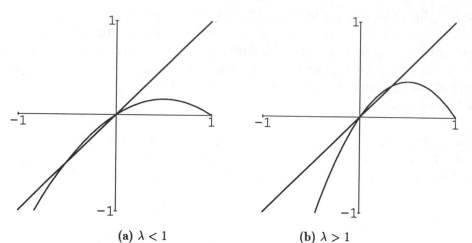

(a) $\lambda < 1$ (b) $\lambda > 1$

Fig. 6.7 A bifurcation in the logistic family $F_\lambda(x) = \lambda x(1-x)$.

that were given by

$$p_\pm(c) = \frac{1}{2}\left(1 \pm \sqrt{1-4c}\right).$$

In the bifurcation diagram we plot the parameter c on the horizontal axis versus the fixed points on the vertical axis. It is perhaps perverse to plot x-values on the vertical axis, but such is life! The bifurcation diagram for Q_c is depicted in Figure 6.8. Note that along each vertical line $c =$ constant, we see either zero, one, or two points corresponding to the fixed points of Q_c. In Figure 6.9 we have sketched the bifurcation diagrams for the bifurcations in the families $e^x + \lambda$ and $\lambda x(1-x)$.

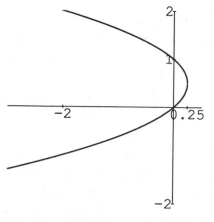

Fig. 6.8 The bifurcation diagram for $Q_c(x) = x^2 + c$.

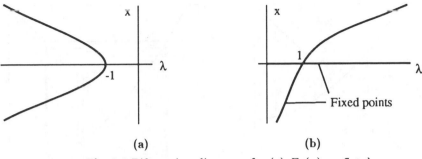

(a)   (b)

Fig. 6.9 Bifurcation diagrams for **(a)** $E_\lambda(x) - e^x + \lambda$
and **(b)** $F_\lambda(x) = \lambda x(1 - x)$ (only for $\lambda > 0$).

6.3   The Period-Doubling Bifurcation

Now we turn our attention to the second of the bifurcations encountered
in the quadratic family $Q_c(x) = x^2 + c$, the period-doubling bifurcation.

Definition. A one-parameter family of functions F_λ undergoes a *period-doubling bifurcation* at the parameter value $\lambda = \lambda_0$ if there is an open interval
I and an $\epsilon > 0$ such that:

1. For each λ in the interval $[\lambda_0 - \epsilon, \lambda_0 + \epsilon]$, there is a unique fixed point
   p_λ for F_λ in I.
2. For $\lambda_0 - \epsilon < \lambda \le \lambda_0$, F_λ has no cycles of period 2 in I and p_λ is
   attracting (resp. repelling).
3. For $\lambda_0 < \lambda < \lambda_0 + \epsilon$, there is a unique 2-cycle q_λ^1, q_λ^2 in I with
   $F_\lambda(q_\lambda^1) = q_\lambda^2$. This 2-cycle is attracting (resp. repelling). Meanwhile,
   the fixed point p_λ is repelling (resp. attracting).
4. As $\lambda \to \lambda_0$, we have $q_\lambda^i \to p_{\lambda_0}$.

Remarks:

1. Thus there are two typical cases for a period-doubling bifurcation. As the
parameter changes, a fixed point may change from attracting to repelling
and, at the same time, give birth to an attracting 2-cycle. Alternatively, the
fixed point may change from repelling to attracting and, at the same time,
give birth to a repelling cycle of period 2.
2. As in the saddle-node case, the direction in which the bifurcation occurs
may be reversed. Also, cycles may undergo a period-doubling bifurcation.
In this case a cycle of period n will give birth to a cycle of period $2n$.

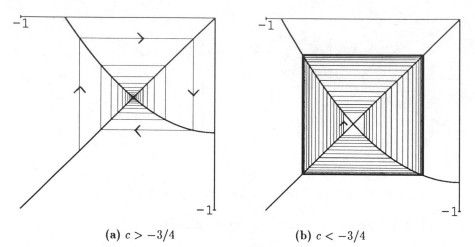

(a) $c > -3/4$ (b) $c < -3/4$

Fig. 6.10 For $c > -3/4$, Q_c has an attracting fixed point, but for $c < -3/4$, this fixed point is repelling and there is an attracting 2-cycle.

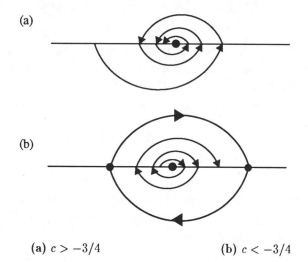

(a) $c > -3/4$ (b) $c < -3/4$

Fig. 6.11 Phase portraits near the period-doubling bifurcation for Q_c.

3. The period-doubling bifurcation occurs when the graph of F_λ is perpendicular to the diagonal, that is, when $F_\lambda'(p_{\lambda_0}) = -1$. By the chain rule, it follows that $(F_\lambda^2)'(p_{\lambda_0}) = 1$, so the graph of the second iterate of F_λ is tangent to the diagonal when the period-doubling bifurcation occurs.*

To understand the period-doubling bifurcation, let's return to the quad-

---

*   The precise conditions that guarantee the occurrence of a period-doubling bifurcation may be found in Devaney, page 90.

ratic family $Q_c(x) = x^2 + c$. We described analytically the period-doubling bifurcation that occurs in this family at $\lambda = -3/4$ in the first section of this chapter. Now let's investigate this bifurcation geometrically. In Figure 6.10 we have performed graphical analysis on Q_c for two distinct c-values, one before the bifurcation and one after. Note how the fixed point p_- changes from attracting to repelling as c decreases through $-3/4$. Also observe how orbits near p_- now tend to the attracting 2-cycle after the bifurcation. In Figure 6.11 we have sketched the phase portraits before and after the bifurcation.

In Figure 6.12 we display the graphs of Q_c^2 for c-values before, at, and after the period-doubling bifurcation. Note how the graph "twists" through the diagonal at the point of bifurcation to produce the 2-cycle. Figure 6.13 depicts the bifurcation diagram for this family. In this picture we have displayed both the fixed points and the 2-cycle for Q_c. Note how the 2-cycle is spawned by the fixed point as c decreases through $-3/4$.

Example. Consider the family $F_\lambda(x) = \lambda x - x^3$. Note that this family has a fixed point at 0 for all λ. When $-1 < \lambda < 1$, this fixed point is attracting. When $\lambda = -1$, 0 is neutral. And when $\lambda < -1$, 0 is repelling. A period doubling bifurcation occurs at $\lambda = -1$. The easiest way to see this is to note that F_λ is an odd function. That is, $F_\lambda(-x) = -F_\lambda(x)$ for all x. In particular, if $F_\lambda(x_0) = -x_0$, then we must have $F_\lambda^2(x_0) = x_0$. This follows from

$$F_\lambda^2(x_0) = F_\lambda(-x_0) = -F_\lambda(x_0) = x_0.$$

Thus we may solve the equation $F_\lambda(x) = -x$ to find 2-cycles. This yields the equation

$$\lambda x - x^3 = -x$$

whose roots are $x = 0, \pm\sqrt{\lambda + 1}$. The point 0 is our fixed point, and the other roots form a 2-cycle. Note that this cycle only appears when $\lambda > -1$. An easy computation shows that the 2-cycle is repelling for these λ-values (See exercise 15). Figure 6.14 displays graphical analysis of F_λ before and after the bifurcation, and Figure 6.15 gives the bifurcation diagram.

6.4   Experiment: The Transition to Chaos

Goal: With an eye toward the story that will begin to unfold in the next chapter, our goal in this experiment will be to observe the dynamical behavior of $Q_c(x) = x^2 + c$ for a large number of c-values in the interval $-2 \leq c \leq 0.25$.

Procedure: You are to compute the orbit of 0 under $Q_c(x) = x^2 + c$ for at least 50 different c-values as specified below. For each such c, the goal is to

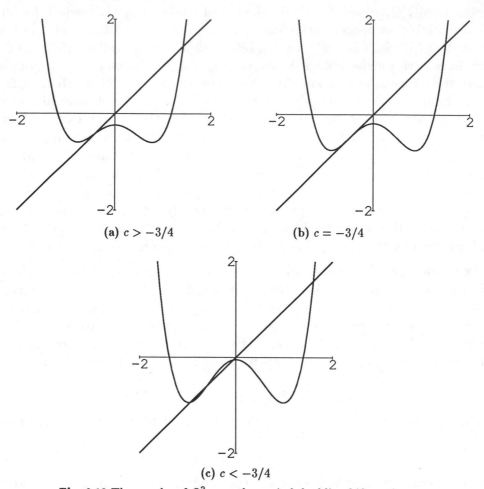

(a) $c > -3/4$

(b) $c = -3/4$

(c) $c < -3/4$

Fig. 6.12 The graphs of Q_c^2 near the period-doubling bifurcation.

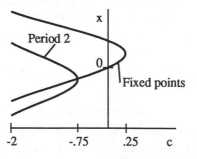

Fig. 6.13 The bifurcation diagram for Q_c.

record the ultimate or "asymptotic" behavior of the orbit of 0. This means that we are only interested in what eventually happens to the orbit of 0. Is it attracted to a fixed point? To a k-cycle? Is there no pattern to the orbit at all?

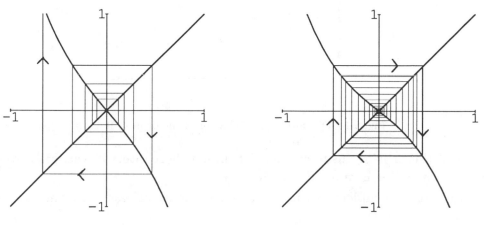

(a) $\lambda < -1$ (b) $-1 < \lambda < 0$

Fig. 6.14 The period-doubling bifurcation for the family $F_\lambda(x) = \lambda x - x^3$.

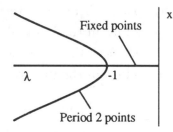

Fig. 6.15 The bifurcation diagram for $\lambda x - x^3$.

For each chosen c you should record both c and the ultimate behavior of the orbit of 0. Then you should plot all of this information as follows. On a piece of paper, plot the c-axis horizontally with $-2 \le c \le 0.25$ and the x-axis vertically with $-2 \le x \le 2$ as depicted in Figure 6.16.

In this diagram, for each chosen c, we will record the ultimate fate of the orbit of 0 as follows. If the orbit of 0 under Q_{c_1} is attracted to a fixed point p_1, then plot (c_1, p_1). If the orbit of 0 under Q_{c_2} is attracted to a 2-cycle, q_1 and q_2, then plot (c_2, q_1) and (c_2, q_2). If there is no pattern to the orbit of 0, then use a histogram to determine approximately the interval (or intervals)

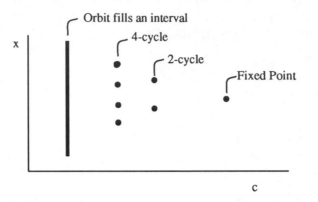

Fig. 6.16 The asymptotic behavior of the orbit of 0.

in which this orbit seems to wander. Then plot these intervals vertically over the corresponding c-value. See Figure 6.16.

To make a complete picture, you should choose at least 50 c-values distributed as follows:

1. Five c's in the interval $-0.75 < c < 0.25$.
2. Five c's in the interval $-1.25 < c < -0.75$.
3. At least twenty equally spaced c's in the interval $-1.4 < c < -1.25$.
4. At least five equally spaced c's in the interval $-1.75 < c < -1.4$.
5. At least ten equally spaced c's in the interval $-1.78 < c < -1.76$.
6. At least five equally spaced c's in the interval $-2 < c < -1.78$.

Notes and Questions:

1. We will see later why we chose the orbit of 0 in this experiment.

2. A similar experiment may be conducted with the family of logistic functions $F_\lambda(x) = \lambda x(1-x)$. Here we plot λ on the horizontal axis with $0 \le \lambda \le 4$ and x on the vertical axis with $0 \le x \le 1$. It is important to use the initial seed $x_0 = 0.5$ in this case, not $x_0 = 0$, which is always fixed for F_λ.

3. For a more complete picture, work may be divided among several groups of experimenters, each assuming responsibility for a different collection of c or λ-values.

Exercises

1. Each of the following functions undergoes a bifurcation of fixed points at the given parameter value. In each case, use algebraic or graphical methods to identify this bifurcation as either a saddle-node or period-doubling bifurcation, or neither of these. In each case, sketch the phase portrait for typical parameter values below, at, and above the bifurcation value.

a. $F_\lambda(x) = x + x^2 + \lambda, \ \lambda = 0$
b. $F_\lambda(x) = x + x^2 + \lambda, \ \lambda = -1$
c. $G_\mu(x) = \mu x + x^3, \ \mu = -1$
d. $G_\mu(x) = \mu x + x^3, \ \mu = 1$
e. $S_\mu(x) = \mu \sin x, \ \mu = 1$
f. $S_\mu(x) = \mu \sin x, \ \mu = -1 \leftarrow 1, c.$
g. $F_c(x) = x^3 + c, \ c = 2/3\sqrt{3}$
h. $E_\lambda(x) = \lambda(e^x - 1), \ \lambda = -1$
i. $E_\lambda(x) = \lambda(e^x - 1), \ \lambda = 1$
j. $H_c(x) = x + cx^2, \ c = 0$
k. $F_c(x) = x + cx^2 + x^3, \ c = 0$

The next four exercises apply to the family $Q_c(x) = x^2 + c$.

2. Verify the formulas for the fixed points $p_\pm$ and the 2-cycle $q_\pm$ given in the text.

3. Prove that the cycle of period 2 given by $q_\pm$ is attracting for $-5/4 < c < -3/4$.

4. Prove that this cycle is neutral for $c = -5/4$.

5. Prove that this cycle is repelling for $c < -5/4$.

Exercises 6–14 deal with the logistic family of functions given by $F_\lambda(x) = \lambda x(1 - x)$.

6. For which values of λ does F_λ have an attracting fixed point at $x = 0$?

7. For which values of λ does F_λ have a nonzero attracting fixed point?

8. Describe the bifurcation that occurs when $\lambda = 1$.

9. Sketch the phase portrait and bifurcation diagram near $\lambda = 1$.

10. Describe the bifurcation that occurs when $\lambda = 3$.

11. Sketch the phase portrait and bifurcation diagram near $\lambda = 3$.

12. Describe the bifurcation that occurs when $\lambda = -1$.

13. Sketch the phase portrait and bifurcation diagram near $\lambda = -1$.

14. Compute an explicit formula for the periodic points of period 2 for F_λ.

15. Consider $F_\lambda(x) = \lambda x - x^3$. Show that the 2-cycle given by $\pm\sqrt{\lambda+1}$ is repelling when $\lambda > -1$.

16. Consider the family of functions $F_\lambda(x) = x^5 - \lambda x^3$. Discuss the bifurcation of 2-cycles that occurs when $\lambda = 2$. *Hint:* Note that this is an odd function for each λ, so points of period 2 may be found by solving $F_\lambda(x) = -x$.

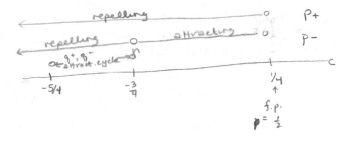

CHAPTER 7

The Quadratic Family

In this chapter we will continue our study of the dynamics of the quadratic functions $Q_c(x) = x^2 + c$. In the previous section we saw that the dynamics of these functions were relatively simple when $c > -5/4$. In this chapter we will jump ahead a bit to study the case where $c \leq -2$. We will see that a remarkable change has occurred—for these values of c the dynamical behavior is much more complicated.

7.1   The Case c = −2

In Experiment 3.6 you computed a variety of orbits of $Q_{-2}(x) = x^2 - 2$ for initial seeds in the interval $-2 \leq x \leq 2$. You undoubtedly saw that almost all orbits behaved in a rather haphazard fashion. In particular, you probably found very few periodic or eventually periodic orbits (with the exception, perhaps, of the fixed points at 2 and −1 and the eventually fixed points at −2, 0, and 1). Here we will see that there are, in fact, many more periodic points for this function that the computer missed.

Recall that all of the interesting dynamics of Q_c take place in the interval $-p_+ \leq x \leq p_+$ where p_+ is the repelling fixed point with $p_+ > 0$. All other orbits of Q_c tend to infinity. For $c = -2$, we have $p_+ = 2$, so we will concentrate on the interval $[-2, 2]$ which, in this section, we will call I. The graph of Q_{-2} on I is shown in Figure 7.1. Note that Q_{-2} is increasing on the interval $[0, 2]$ and takes this subinterval onto the entire interval I in one-to-one fashion. Similarly, Q_{-2} is decreasing on the subinterval $[-2, 0]$ and also takes this interval onto I in one-to-one fashion. Thus every point (with the sole exception of −2) in I has exactly two preimages in I: one in $[-2, 0]$ and the other in $[0, 2]$. Geometrically, you should think of Q_{-2} as first stretching and folding the interval I and then mapping it back over itself twice as depicted in Figure 7.2.

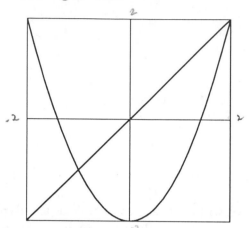

[handwritten: $Q_{-2}(x) = x^2 - 2$]

[handwritten: $I = [-2, 2]$ in this chapter]

Fig. 7.1 The graph of Q_{-2} on the interval $[-2, 2]$.

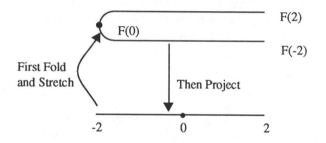

Fig. 7.2 Q_{-2} folds I over itself so that each point (except -2) is covered twice.

Now let's turn to the graph of Q_{-2}^2. Since one application of Q_{-2} takes the interval $[0, 2]$ onto I, the second iteration of Q_{-2} on this interval should fold this interval over all of I exactly as Q_{-2} did on I. That is, with the exception of -2, all points in I should have two preimages in $[0, 2]$ under Q_{-2}^2. Of course, the same happens in the interval $[-2, 0]$. This is the reason why the graph of Q_{-2}^2 has exactly two "valleys" as shown in Figure 7.3a. Arguing similarly, there are four subintervals of I that are mapped by Q_{-2}^2 exactly onto I, and hence Q_{-2}^3 must fold these intervals over I as before. So the graph of Q_{-2}^3 has four valleys in I as shown in Figure 7.3b.

Continuing in this fashion, we see that the graph of Q_{-2}^n has 2^{n-1} valleys in the interval I. Hence this graph must cross the diagonal at least 2^n times over I. Later we will see that this graph crosses the diagonal exactly 2^n times.

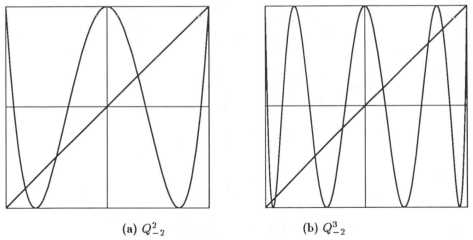

(a) Q_{-2}^2 (b) Q_{-2}^3

Fig. 7.3 The graphs of higher iterates of Q_{-2} on $[-2, 2]$.

In any event, each crossing gives us a periodic point of (not necessarily prime) period n, so we have proved the following

Theorem. *The function Q_{-2} has at least 2^n periodic points of period n in the interval $-2 \le x \le 2$.*

You should contrast this with the results we saw in Experiment 3.6. With the computer you found very few periodic points for Q_{-2}, yet we now know that there are infinitely many of them. Clearly, more is going on for this function than meets the eye.

This also raises an important question. For the quadratic family Q_c, for c-values larger than $-5/4$, we saw that there were very few periodic points. On the other hand, by the time c reaches -2, infinitely many periodic points of all periods have arisen. The big question is: How did these periodic points arise? Where did they come from? This too is one of the main questions we will address in the succeeding chapters.

7.2 The Case c < −2

In Experiment 3.6 you also computed a variety of orbits of Q_c when $c < -2$. You most likely saw that nearly all orbits of Q_c tended to infinity in this case, so it appears that the dynamics in this case are relatively simple. As in the previous section, we will see now that this is by no means the case.

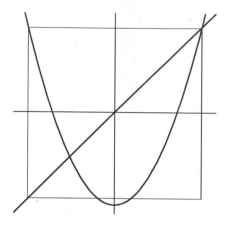

Fig. 7.4 The graph of Q_c for $c < -2$.

For the remainder of this section we will fix a particular c-value with $c < -2$. It is important to note that many of the intervals described below depend on which c-value is chosen, so we really should index everything with a c. However, to keep the notation simple, we will avoid this.

In Figure 7.4 we show the graph of Q_c for a c-value less than -2. In this figure we also show a box whose vertices include the points $(-p_+, -p_+)$ and (p_+, p_+). Note that the graph of Q_c protrudes through the bottom of this box, unlike the previous cases. As before, we denote the interval $-p_+ \le x \le p_+$ by I. Note that, unlike the case $c = -2$ where the point $x = -2$ had only one preimage in I, here all points in I have exactly two preimages in I. Also, graphical analysis shows that there is an open interval in I containing 0 that is mapped outside of I by Q_c. This interval is precisely the set of points lying above the portion of the graph outside of the box. In particular, the orbit of any point in this interval escapes from I and so tends to infinity. Let's call this interval A_1. A_1 is the set of points that escape from I after just one iteration of Q_c (see Fig. 7.5).

Now any orbit that eventually leaves I must tend to infinity, so we understand the fate of all these orbits. Therefore it remains to understand the fate of orbits that never escape from I. Let us denote this set of points by Λ. That is,

$$\Lambda = \{x \in I \mid Q_c^n(x) \in I \text{ for all } n\}.$$

The set Λ contains all of the interesting orbits of Q_c. The first question is: What exactly is this set Λ? What does it look like? To answer this, we will describe instead the complement of Λ. From Fig 7.6a, we see that there are a pair of open intervals that have the property that, if x is in one of these

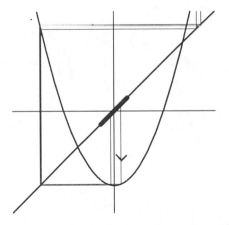

Fig. 7.5 Any orbit in the interval A_1 tends to infinity.

intervals, then $Q_c(x) \in A_1$. Hence the orbit of x escapes from I after two iterations. We call this pair of intervals A_2. So if $x \in A_2$, it follows that $Q_c^n(x) \to \infty$ and $x \notin \Lambda$. Continuing, from Figure 7.6b, we see that there are four smaller open subintervals having the property that if x belongs to one of these intervals, then $Q_c(x) \in A_2$ and again the orbit of x escapes. We denote the union of these intervals by A_3. So A_3 is the set of points that escape from I after three iterations. Continuing in this fashion, let A_n denote the set of points in I whose orbit leaves I after exactly n iterations. Since each point in I has exactly two preimages in I, the set A_n consists of exactly 2^{n-1} open intervals. If a point has an orbit that eventually escapes from I, then this point must lie in A_n for some n. Hence the complement of Λ in I is just the union of the A_n for all n.

What does Λ look like? To construct Λ, we first remove A_1 from the interval I, leaving two closed intervals behind. Then we throw out A_2, which from Figure 7.6 leaves us with four closed intervals. Then we remove A_3, leaving eight closed intervals. Continuing, when we remove A_n, we are left with 2^n closed intervals. The set Λ is what is left after we remove all of these open intervals. This set is called a *Cantor set*.

Two questions immediately arise. First, are there any points left after we throw out all of these intervals? The answer is yes, for the fixed points at $p_\pm$ certainly lie in Λ. Furthermore, the endpoints of each interval that remains after removing the A_n are also in Λ, for these points are eventually fixed (after n iterations they land on $-p_+$ and so after $n+1$ iterations they land on the fixed point p_+). We will see later that in fact there are many, many more points in this set.

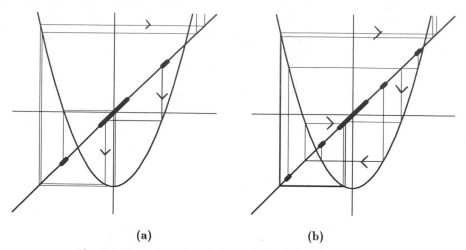

(a)   (b)

Fig. 7.6 The intervals (a) A_2 and A_1, (b) and A_3 and A_1.

A second observation is that Λ contains no intervals. This is a little more difficult to see. In fact we will make a simplifying assumption that allows us to prove this more easily. Let us assume that $c < -(5+2\sqrt{5})/4 = -2.368\ldots$. The effect of this assumption is that $|Q_c'(x)| > 1$ for all $x \in I - A_1$. In Exercise 8 you are asked to verify this fact. In particular, there is a constant $\mu > 1$ such that $|Q_c'(x)| \geq \mu$ for all $x \in I - A_1$.*

Now suppose that Λ contains an interval J. Let's say that J has length $\ell > 0$. Since $|Q_c'(x)| > \mu$ for all $x \in J$, it follows from the Mean Value Theorem that the length of $Q_c(J)$ exceeds $\mu\ell$. This is true since, for any two points x and y in J, the Mean Value Theorem shows that

$$|Q_c(x) - Q_c(y)| > \mu|x - y|$$

(see section 5.4). Also, since $J \subset \Lambda$, we also have $Q_c(J) \subset \Lambda$. Hence we may apply the same argument to the interval $Q_c(J)$, and we therefore conclude that the length of $Q_c^2(J)$ exceeds $\mu \cdot$ length of $Q_c(J) > \mu^2\ell$. Continuing, we find that the length of $Q_c^n(J)$ is larger than $\mu^n\ell$. But $\mu > 1$ so $\mu^n \to \infty$. Therefore, the length of $Q_c^n(J)$ becomes arbitrarily large. However, the interval $Q_c^n(J)$ must lie within I, which has finite length. This contradiction establishes the result.

Finally, we observe that Λ is a closed subset of I. Indeed, the complement of Λ consists of the open intervals $(-\infty, -p_+)$, (p_+, ∞), and all of the A_n.

---

* Just let $\mu = Q_c'(x_1)$ where x_1 is the right-hand endpoint of A_1. Since $Q_c'' > 0$, we have $Q_c'(x) > Q_c'(x_1)$ if $x > x_1$.

Since a the union of a collection of open intervals is open, it follows that the complement of Λ is an open set and so Λ is closed. See Appendix A.3 for more details on open and closed sets. Let's summarize all of these facts:

Theorem. *Suppose $c < -2$. Then the set of points Λ whose orbits under Q_c do not tend to infinity is a nonempty closed set in I that contains no intervals.*

Remarks:
1. This theorem is valid for all c-values less than -2. Our proof only works for $c < -(5 + 2\sqrt{5})/4 = -2.368\ldots$. The case of c-values closer to -2 is considerably more difficult and will be omitted.
2. A subset of $\mathbf{R}$ that contains no intervals is called *totally disconnected*. We will meet many such sets in the sequel, including the next section where we investigate the prototype of such sets, the *Cantor middle-thirds set*.

7.3   The Cantor Middle-Thirds Set

In this section we will discuss the construction of the classical *Cantor middle-thirds set*, or Cantor set for short. While this set may seem quite "pathological" at first, we will see that these kinds of sets arise over and over again in dynamics. Moreover, the Cantor set is the most basic kind of *fractal,* a topic we will investigate in detail in Chapter 14.

To construct this set we begin with the interval $0 \leq x \leq 1$. From this we remove the middle third: the open interval $\frac{1}{3} < x < \frac{2}{3}$. Note that we leave the endpoints behind. What remains is a pair of closed intervals, each of them one-third as long as the original. Now do this again: From each of the remaining intervals we remove the middle third. That is, from the interval $0 \leq x \leq \frac{1}{3}$, we remove $\frac{1}{9} < x < \frac{2}{9}$, and from $\frac{2}{3} \leq x \leq 1$ we remove $\frac{7}{9} < x < \frac{8}{9}$. Note that, in each case, we have again left behind the endpoints; now four closed intervals remain, each one-ninth as long as the original interval.

We now repeat this process over and over. At each stage, we remove the middle third of each of the intervals remaining at the previous stage. What is left in the limit is the Cantor middle-thirds set, which we denote by K. See Figure 7.7.

Note how similar this construction is to the construction of the set Λ in the previous section. Indeed, it is easy to check that K is closed (same argument as before) and totally disconnected. Certainly there are a lot of

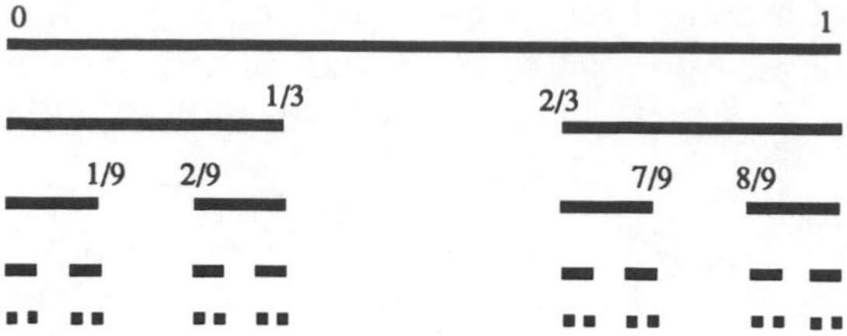

Fig. 7.7 Construction of the Cantor middle-thirds set.

points in K; any endpoint of one of the removed middle-thirds intervals is in K. One of the remarkable features of the Cantor set is the fact that there are many, many more points in K. In fact, K is an *uncountable set.**

To explain this remark, we recall that a set is called *countable* if it can be put in one-to-one correspondence with the set of positive integers, that is, we can enumerate the points in the set. A set is *uncountable* if it is neither finite nor countable. For example, the set of rational numbers is a countable set whereas the set of all real numbers, the set of all irrationals, and any interval (a, b) with $a \neq b$ are uncountable. In general, countable sets are "small" compared to uncountable sets.

To see that the Cantor set is uncountable, we make a brief digression into ternary expansions of real numbers. Recall that the geometric series

$$\sum_{i=0}^{\infty} a^i = 1 + a + a^2 + a^3 + \cdots$$

converges absolutely if $|a| < 1$ and, in fact,

$$\sum_{i=0}^{\infty} a^i = \frac{1}{1-a}.$$

More generally,

$$\sum_{i=k}^{\infty} a^i = \frac{a^k}{1-a}.$$

For example,

$$\sum_{i=0}^{\infty} \left(\frac{2}{3}\right)^i = 3$$

and

---

*   For more information on countable and uncountable sets, refer to W. Rudin, *Principles of Mathematical Analysis* New York: McGraw-Hill, 1964.

$$\sum_{i=1}^{\infty} \left(\frac{1}{3}\right)^i = \left(\frac{\frac{1}{3}}{1 - \frac{1}{3}}\right) = \frac{1}{2}.$$

Now suppose for each positive integer i, s_i is either 0, 1, or 2. Then the series

$$\sum_{i=1}^{\infty} \frac{s_i}{3^i}$$

is dominated by the convergent geometric series

$$\sum_{i=1}^{\infty} \frac{2}{3^i} = 2 \left(\frac{\frac{1}{3}}{1 - \frac{1}{3}}\right) = 1$$

in the sense that

$$0 \le \frac{s_i}{3^i} \le \frac{2}{3^i}$$

for each i. Hence, by the Comparison Test, this series converges and we have

$$0 \le \sum_{i=1}^{\infty} \frac{s_i}{3^i} \le 1.$$

Definition. The sequence of integers $0.s_1 s_2 s_3 \ldots$ where each s_i is either 0,1, or 2 is called the *ternary expansion* of x if

$$x = \sum_{i=1}^{\infty} \frac{s_i}{3^i}.$$

Example. The sequence $0.020202\ldots$ is the ternary expansion of $1/4$ since

$$\frac{0}{3^1} + \frac{2}{3^2} + \frac{0}{3^3} + \frac{2}{3^4} + \cdots = 2 \left[\sum_{i=1}^{\infty} \left(\frac{1}{9}\right)^i\right] = \frac{1}{4}.$$

The sequence $0.012012012\ldots$ is the ternary expansion for $5/26$ since

$$\frac{0}{3^1} + \frac{1}{3^2} + \frac{2}{3^3} + \frac{0}{3^4} + \frac{1}{3^5} + \frac{2}{3^6} + \cdots = \frac{1}{3^2} + \frac{1}{3^5} + \cdots + \frac{2}{3^3} + \frac{2}{3^6} + \cdots$$

$$= \frac{1}{9} \sum_{i=0}^{\infty} \left(\frac{1}{27}\right)^i + \frac{2}{27} \sum_{i=0}^{\infty} \left(\frac{1}{27}\right)^i = \frac{5}{26}.$$

Note that certain x-values in $[0, 1]$ may have two different ternary expansions. For example, $1/3$ has expansions $0.10000\ldots$ and $0.02222\ldots$ since

$$\sum_{i=2}^{\infty} \frac{2}{3^i} = \frac{2}{9} \sum_{i=0}^{\infty} \left(\frac{1}{3}\right)^i = \frac{1}{3}.$$

Similarly, $8/9$ has expansion $0.220000\ldots$ as well as $0.212222\ldots$ since

$$\frac{2}{3} + \frac{1}{9} + \sum_{i=3}^{\infty} \frac{2}{3^i} = \frac{7}{9} + \frac{2}{27} \sum_{i=0}^{\infty} \frac{1}{3^i} = \frac{8}{9}.$$

This is similar to the problem that occurs for binary expansions where sequences of 0's and 1's of the form $0.s_1 \ldots s_n 10000\ldots$ and $0.s_1 \ldots s_n 01111\ldots$ represent the same numbers. For ternary expansions, it is easy to check that the sequences of the form $0.s_1 \ldots s_n 10000\ldots$ and $0.s_1 \ldots s_n 02222\ldots$ represent the same number. Also, the two sequences of the form $0.s_1 \ldots s_n 20000\ldots$ and $0.s_1 \ldots s_n 12222\ldots$ correspond to the same number. Note that the numbers for which these ambiguities occur are precisely the numbers with finite ternary expansion $0.s_1 \ldots s_k 0000\ldots$ for some k which in turn are precisely the rational numbers that may be written in the form $\frac{p}{3^k}$ for some integer p with $0 \leq p < 3^k$.

How do we find the ternary expansion of a number? We will see an easy method later that involves dynamics, but for now we simply note that if x has ternary expansion $0.s_1 s_2 \ldots$, then the digit s_1 determines which third of the interval $[0, 1]$ x lies in. If $s_1 = 0$, then $x \in [0, \frac{1}{3}]$; if $s_1 = 1$, then $x \in [\frac{1}{3}, \frac{2}{3}]$; and if $s_1 = 2$, then $x \in [\frac{2}{3}, 1]$. The reason for this is that the "tail" of the series representing x, namely

$$\sum_{i=2}^{\infty} \frac{s_i}{3^i}$$

is no larger than

$$\sum_{i=2}^{\infty} \frac{2}{3^i} = \frac{1}{3}.$$

Notice that the ambiguous x-values are exactly those that share two ternary representations. So given the digit s_1, we know which third of the unit interval x lies in: the left third if $s_1 = 0$, the middle third if $s_1 = 1$, and the right third if $s_1 = 2$.

Arguing exactly as above, we see that the second digit s_2 tells us which third of these subintervals x lies in. That is, s_2 determines the left, middle, or right third of the interval determined at the previous stage. As above, the reason is that the series

$$\sum_{i=3}^{\infty} \frac{s_i}{3^i},$$

which represents the tail of the ternary expansion, is in this case no larger than 1/9. Continuing in this fashion, we see that ternary expansions have a direct relationship to points in the Cantor set. In particular, if x has ternary expansion for which some entry $s_n = 1$, then x must lie in one of the middle-thirds intervals that was removed during the construction of K. The only exception to this is if x is an endpoint of this interval, in which case x has an alternative ternary expansion that has no 1's whatsoever. Thus we may say that the Cantor set is the set of real numbers in $[0,1]$ for which there is a ternary expansion containing no 1's.

We now see why the Cantor set is uncountable. We can set up a correspondence between points in the Cantor set and points in the interval $[0,1]$ as follows. Given $x \in K$, consider the ternary expansion of x that contains no 1's. Then change every "2" in this expansion to a "1" and consider the resulting sequence as the binary expansion of a number in $[0,1]$. For example, the ternary expansion of 1/4 is 0.020202... as we saw above. So $1/4 \in K$. We now associate to 1/4 the binary sequence 0.010101 ... which corresponds to 1/3 since

$$\sum_{i=1}^{\infty} \left(\frac{1}{4}\right)^i = \frac{1}{3}.$$

The correspondence between binary and ternary sequences is clearly one-to-one since we can go backwards just as easily: given a binary expansion of a number in $[0,1]$, change all the 1's to 2's to get the ternary expansion of a point in K. Thus we have proved the following

Theorem. *The Cantor middle-thirds set is uncountable.*

A natural question is whether or not the Cantor middle-thirds set and our dynamically defined set Λ for the quadratic map are essentially the same, or are *homeomorphic*. (See Appendix A.1.) We will see that they are, but we will use a more powerful tool to show this. This tool is *symbolic dynamics*, which we will discuss in Chapter 9.

Exercises

1. List the intervals that are removed in the third and fourth stages of the construction of the Cantor middle-thirds set.

2. Compute the sum of the lengths of all of the intervals that are removed from the interval $[0,1]$ in the construction of the Cantor middle-thirds set.

In the next five exercises, find the rational numbers whose ternary expansion is given by:

3. $0.2121\overline{21}$

4. $0.022022\overline{022}$

5. $0.0022\overline{2}$

6. $0.212010\overline{1}$

7. $0.0101101\overline{101}$

8. Let $Q_c(x) = x^2 + c$ with $c < -2$. Let I and A_1 be defined as in this chapter. Prove that if $c < -(5+2\sqrt{5})/4$, then $|Q_c'(x)| > 1$ for all $x \in I - A_1$. *Hint:* Note that $|Q_c'(x)| > 1$ for all $x > 1/2$, so it suffices to find the c-values for which $Q_c(1/2) < -p_+$.

Dynamics on the Cantor middle-thirds set: The following exercises deal with the function

$$T(x) = \begin{cases} 3x & \text{if } x \leq 1/2 \\ 3 - 3x & \text{if } x > 1/2 \end{cases}.$$

9. Sketch the graph of T and show by graphical analysis that, if $x > 1$ or $x < 0$, then $T^n(x) \to -\infty$ as $n \to \infty$.

10. Find the fixed points for T. What is the ternary expansion of these points?

11. Show that $3/13$ and $3/28$ lie on 3-cycles for T.

12. Show that if $x \in (1/3, 2/3)$, then $T^n(x) \to -\infty$ as $n \to \infty$.

13. Show that if $x \in (1/9, 2/9)$ or $x \in (7/9, 8/9)$, then $T^n(x) \to -\infty$ as $n \to \infty$.

14. Let $\Gamma = \{x \in [0,1] \,|\, T^n(x) \in [0,1] \text{ for all } n\}$. Prove that $\Gamma = K$, the Cantor middle-thirds set.

15. Suppose $x \in \Gamma$ has ternary expansion $0.a_1a_2a_3\ldots$. What is the ternary expansion of $T(x)$? Be careful: there are two very different cases!

The Logistic Map: The following exercises deal with the logistic map $F_\lambda(x) = \lambda x(1 - x)$.

16. Prove that F_4 has at least 2^n periodic points of (not necessarily prime) period n in the interval $[0, 1]$.

17. Prove that if $x > 1$ or $x < 0$, then $F_4^n(x) \to -\infty$ as $n \to \infty$.

18. Prove that if $x \in [0, 1]$, $F_4^n(x) \not\to -\infty$ as $n \to \infty$.

19. In an essay, describe the dynamics of F_λ when $\lambda > 4$. In the essay describe the set of points whose orbit tends to $-\infty$. Can you find a λ-value for which $|F_\lambda'(x)| > 1$ for all x whose orbit does not escape?

20. Consider the function $S(x) = \pi \sin x$. Prove that S has at least 2^n periodic points of period n in the interval $[0, \pi]$.

21. Show that the function $T(x) = x^3 - 3x$ has at least 3^n periodic points of period n in the interval $[-2, 2]$.

CHAPTER 8

Transition to Chaos

In previous chapters we have investigated the dynamics of $Q_c(x) = x^2 + c$ at length. We have seen that for c-values above $-5/4$ the dynamics of this function are quite tame. On the other hand, by the time c reaches -2, the dynamics become much more complicated. Our goal in this and the next few sections is to understand how Q_c and other similar functions make the transition from simple to complicated dynamics.

8.1   The Orbit Diagram

In this section, all of our work will be experimental. We will make a series of observations based on computer images. We will spend much of the rest of this book trying to explain what we have seen here.

We will deal here with the *orbit diagram*. This image is one of the most instructive and intricate images in all of dynamical systems. The orbit diagram is an attempt to capture the dynamics of Q_c for many different c-values in one picture. The result gives a good summary of the dynamics of the entire family as well as an idea of how Q_c makes the transition to chaos.

In the orbit diagram we plot the parameter c on the horizontal axis versus the *asymptotic orbit* of 0 under Q_c on the vertical axis. By asymptotic orbit we simply mean that we do not plot the first few iterations (usually 100 or so) of 0. This allows the orbit to settle down and reach its eventual behavior; thus we can see the fate of the orbit. By not plotting the first few iterations, we have eliminated the "transient behavior" of the orbit.

For the orbit diagram of Q_c we initially choose the parameter c in the range $-2 \leq c \leq 1/4$ since, as we will see in the next two chapters, we understand the dynamics of Q_c completely when c is outside this interval. We plot the asymptotic orbit of 0 under Q_c on the vertical line over c. As we know, the orbit of 0 remains in the interval $[-2, 2]$ for these c-values, so the coordinates on the vertical axis run from -2 to 2. The program to generate the orbit diagram of Q_c is listed in Appendix B. Incidentally, this is the picture you computed by hand in Experiment 6.4.

A natural question is why do we use the orbit of 0 to plot the orbit diagram. We will see the real answer to this question in Chapter 12. For now, we simply note that 0 is the only critical point of Q_c.

Definition. Suppose $F : \mathbf{R} \to \mathbf{R}$. A point x_0 is a *critical point* of F if $F'(x_0) = 0$. The critical point x_0 is *nondegenerate* if $F''(x_0) \neq 0$. Otherwise, the critical point is *degenerate*.

For example, 0 is the only critical point for Q_c, and it is nondegenerate. Similarly, $1/2$ is the only critical point of the logistic function $F_\lambda(x) = \lambda x(1 - x)$ (provided $\lambda \neq 0$), and it too is nondegenerate. On the other hand, $F(x) = x^3$ has a degenerate critical point at 0, as does x^4.

In Figure 8.1 we have displayed the full orbit diagram of Q_c. Note that for c in the range $-3/4 \leq c \leq 1/4$, we see exactly one point on the vertical line over c. We know why this happens: for these c-values, the orbit of 0 is attracted to an attracting fixed point, and this single point corresponds to this fixed point. Note that this portion of the orbit diagram is exactly the same as the bifurcation diagram introduced earlier. As c decreases below $-3/4$ we see a period-doubling bifurcation: a new attracting cycle of period 2 is born and the two points in the orbit diagram above these points correspond to this cycle. Note that in the orbit diagram we no longer see the repelling fixed point as we do in the bifurcation diagram. This is one of the main differences between these two images. Continuing to decrease c, we seem to see a succession of period doublings. In Figure 8.2 the portion of the orbit diagram inside the rectangular box in Figure 8.1 is magnified, and Figures 8.3 and 8.4 are successive magnifications of the rectangles in the previous figures. This leads to our first observation.

Observation 1: *As c decreases, we seem to see a succession of period-doubling bifurcations. It seems that periodic points first appear in the order* $1, 2, 4, 8, \ldots, 2^n, \ldots$.

Note the large vertical white region far to the left in Figure 8.1. If you look closely, you see that this region is not completely white but rather is

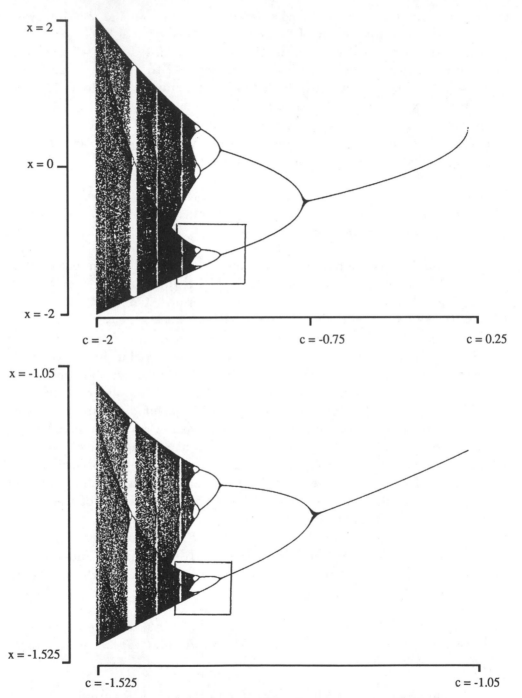

Figs. 8.1, 2 Orbit diagram for $Q_c(x) = x^2 + c$ and a magnification.

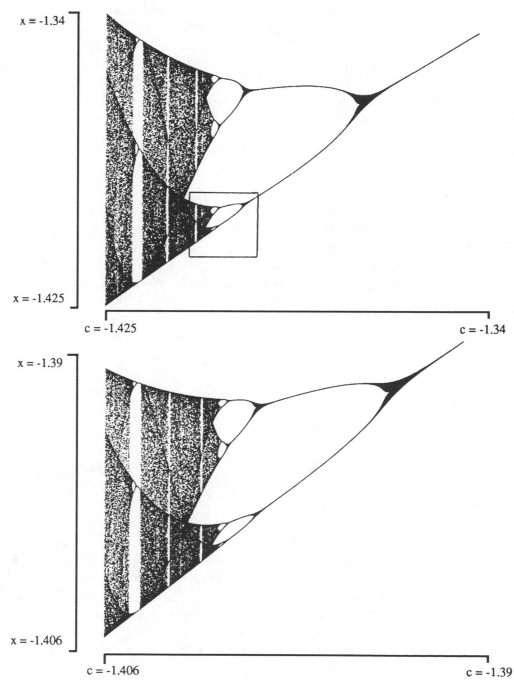

x = -1.34

x = -1.425

c = -1.425

c = -1.34

x = -1.39

x = -1.406

c = -1.406

c = -1.39

Figs. 8.3, 4 Further successive magnifications of the orbit diagram of $Q_c(x) = x^2 + c$.

crossed by three tiny black regions at the top, middle, and bottom. For this reason, this region is called the *period*-3 *window*. In Figure 8.5 this window is magnified. We see clearly what appears to be an attracting 3-cycle, which then seems to undergo a sequence of period-doubling bifurcations. Figure 8.6 shows the central portion of this window magnified to make the period-doublings plainly visible. There are other windows visible in all of these figures. Magnifications of these windows always yield the same surprising result: we seem to see an initial periodic orbit of some period n followed by a succession of period-doubling bifurcations. We call these regions period-n windows.

Observation 2: *In each period-n window, we seem to see the appearance of an attracting n-cycle followed by a succession of period-doubling bifurcations.*

One very obvious feature of the magnifications in Figures 8.1–4 is the similarity of these images. Neglecting for the moment the fact that the single curve in the right of each picture represents a periodic point of period 1, 2, 4, and 8 respectively, we see that the images in these successive magnifications are qualitatively the same. Note the appearance of a "period-3 window" to the left in each image.

Observation 3: *The orbit diagram appears to be self-similar: when we magnify certain portions of the picture, the resulting image bears a striking resemblance to the original figure.*

Of course, the words "striking resemblance" in this observation are not very precise in a mathematical sense. We will have to describe this notion more fully later. But for now, there is no question that the orbit diagram possesses a striking amount of self-similarity.

Note that the curves connecting what appear to be periodic orbits are continuous as c varies—these curves never suddenly jump indicating the co-existence of another attracting cycle. That is, for each c-value, it appears that there is at most one attracting cycle. Of course, since we only compute one orbit for each c-value, there may be other attracting orbits present that we do not see since 0 is not attracted to them. Nevertheless, it is interesting that no jumps occur among the periodic cycles that we do see.

Observation 4: *It appears that there is at most one attracting cycle for each Q_c.*

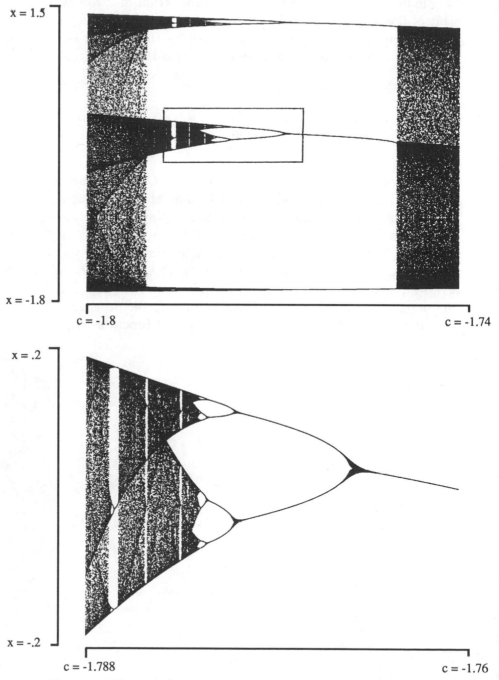

Figs. 8.5, 6 The period 3 window for Q_c and a magnification.

Finally, note that there is a large proportion of c-values for which the orbit of 0 does not seem to settle down on an attracting cycle. Even if we increase the number of iterations as well as the number of iterations not plotted, this regime remains quite large. This is another glimpse of chaotic behavior.

Observation 5: *It appears that there is a large set of c-values for which the orbit of* 0 *is not attracted to an attracting cycle.*

It may appear that the orbit diagram for Q_c is a rather special object. For other functions, we might expect a completely different image when we perform the same experiment. Figure 8.7 displays the orbit diagram for the logistic family $F_\lambda(x) = \lambda x(1 - x)$ with $1 \leq \lambda \leq 4$ and $0 \leq x \leq 1$. For this family we have again plotted the parameter versus the asymptotic orbit of the critical point, which in this case is $1/2$. Although the whole image is reversed, note how similar this diagram is to the diagram for the quadratic function. In the exercises we ask you to modify the program in Appendix B so that the orbit diagrams for other families of functions can be plotted.

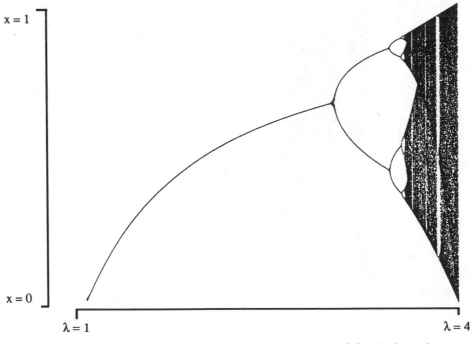

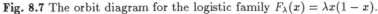

Fig. 8.7 The orbit diagram for the logistic family $F_\lambda(x) = \lambda x(1 - x)$.

8.2   The Period-Doubling Route to Chaos

Our goal in this section is to give a geometric reason for seeing an initial sequence of period doublings in the orbit diagram. We will give a more rigorous explanation of this in Chapter 11 when we describe Sarkovskii's Theorem. Our methods here, involving successive magnifications of the graphs of $Q_c^{2^n}$, come close to the ideas of renormalization group analysis from theoretical physics.

In Figure 8.8 are plotted the graphs of Q_c for six different c-values. In each case we have superimposed a square on the graph with vertices at (p_+, p_+) and $(-p_+, -p_+)$. The six different c-values chosen exhibit dynamical behavior that we have already studied, namely:

   a. Saddle-node bifurcation point.
   b. Critical point is fixed.
   c. Period-doubling bifurcation point.
   d. Critical point has period 2.
   e. Chaotic c-value.
   f. Cantor set case.

Figure 8.9 shows the graphs of Q_c^2 for six different c-values. These values are not the same as those in Figure 8.8; however, note the portions of the graphs inside the small boxes. They resemble very closely the corresponding portions of the graph of Q_c in Figure 8.8, only on a much smaller interval. To be precise, recall that the left-hand fixed point of Q_c is denoted p_-. From the graph of Q_c^2, we see that there is a point to the left of p_- that is mapped by Q_c^2 to p_-. Call this point q. This is the point that lies on the x-axis directly above the left-hand edge of the boxes in Figure 8.9. Then Figures 8.8–9 show that there is a resemblance between the graphs of Q_c on the interval $[-p_+, p_+]$ and those of Q_c^2 on $[q, p_-]$. In particular, we would expect the function Q_c^2 to undergo a similar sequence of dynamical behaviors on this interval as Q_c did on the larger interval. Comparing the small portion of the orbit diagram in Figure 8.2 to the whole diagram in Figure 8.1, we see that this indeed appears to be the case.

Now consider the graphs of Q_c^2 restricted to $[q, p_-]$. Since these graphs each resemble one of the quadratic functions, we expect its second iterate on the small interval to look like the second iterate of one of the original quadratic functions on $[-p_+, p_+]$. In particular, there should be a small subinterval of $[q, p_-]$ on which Q_c^4 looks like Q_c. As c decreases, we would then expect Q_c^4 to undergo the same sequence of bifurcations on this smaller interval as Q_c^2 and Q_c on the larger intervals. Figure 8.3 bears this out.

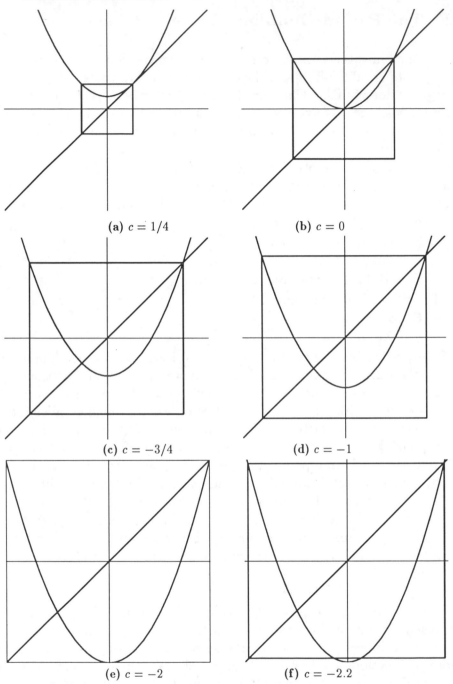

(a) $c = 1/4$ **(b)** $c = 0$

(c) $c = -3/4$ **(d)** $c = -1$

(e) $c = -2$ **(f)** $c = -2.2$

Fig. 8.8 Graphs of Q_c.

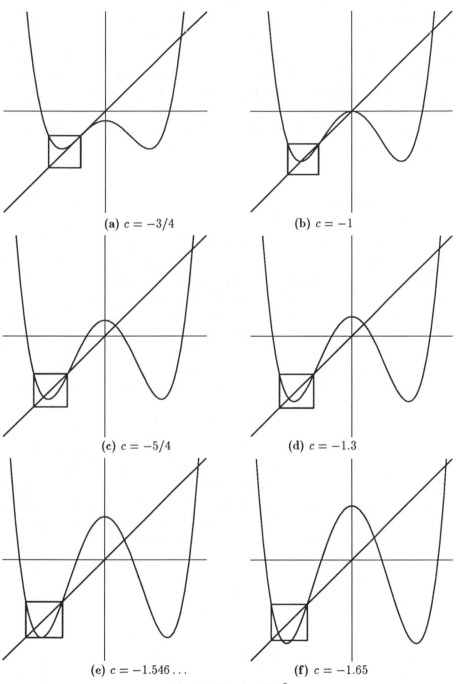

(a) $c = -3/4$

(b) $c = -1$

(c) $c = -5/4$

(d) $c = -1.3$

(e) $c = -1.546\ldots$

(f) $c = -1.65$

Fig. 8.9 Graphs of Q_c^2.

This is the beginning of the process called *renormalization*. At each stage of the process, we consider the second iterate of the map at the previous stage and then zoom in on a small subinterval on which the map resembles a quadratic function. At the nth stage, we find a tiny subinterval on which $Q_c^{2^n}$ resembles the original function. In particular, as c decreases, the graphs of the $Q_c^{2^n}$ make the transition from a saddle-node bifurcation (actually, this is a period-doubling bifurcation viewed from one side), through a period doubling, and on into the chaotic regime. This accounts for the fact that the orbit diagram features regions that are apparently self-similar.

8.3   Experiment: Windows in the Orbit Diagram

Goal: The goal of this experiment is twofold. The first object is for you to see the remarkable structures and patterns that occur in the orbit diagram. The second aim is to investigate the similarities and differences between two typical orbit diagrams, for the quadratic function $Q_c(x) = x^2 + c$ and the logistic function $F_\lambda(x) = \lambda x(1 - x)$. Recall that a period k window in the orbit diagram consists of an interval of parameter values for which we find an initial attracting period k cycle together with all of its period-doubling "descendants." For example, in the big picture of the quadratic family's orbit diagram, we clearly see a period 1 window and a period 3 window. Remember, "period" here means the period of the attracting orbit *before* it begins to period double. See Figure 8.10.

Procedure: The object of this experiment is to catalog a number of smaller windows in the orbit diagrams of both $Q_c(x) = x^2 + c$ and $F_\lambda(x) = \lambda x(1-x)$. Clearly, the period-1 and -3 windows are the most visible for each of the two families. Now magnify the region between these two windows for the quadratic family. What do you see? The two largest windows are clearly of period 5 and 6. This is the "second generation" of windows. There are other smaller windows visible, but we record only the two widest windows at this time. We will record this as follows:

Quadratic function:

Generation 1: 3   1
Generation 2: 3   5   6   1

Note that we put these windows in the right order. Now go to the next generation. Find the periods of the two "largest" windows between the

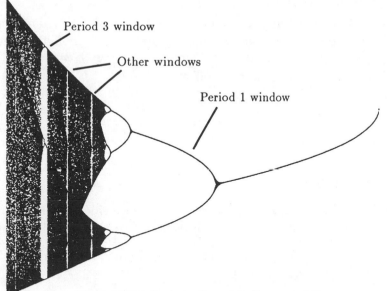

Fig. 8.10 Windows in the orbit diagram of Q_c.

period-1 and -6 window. Call these periods A and B. Then find the periods of the largest windows between the period-5 and -6 windows. Between the period-3 and -5 windows, there should only be one largest window, so find its period. You should thus be able to fill in this chart by supplying the periods A–E, which represent the periods of the largest windows in the order given.

Quadratic function:

Generation 3: 3   E   5   D   C   6   B   A   1

Now proceed to generation 4. In this case we will find only the largest windows between the period-6 and period -1 windows found in the third generation, including the windows between period 6 and period B, period B and period A, and finally period A and period 1. The largest windows have periods α-ϵ as indicated in the following chart:

Quadratic function:

Generation 4: 6 ϵ B δ γ A β α 1

Notes and Questions:

1. The periods of the windows near the left end of this chart may be a judgment call: you don't need to spend hours trying to figure out which window is biggest. Just eyeball it!. Also, you may have to magnify the individual windows in order to read off the periods.

2. Do this for both $Q_c(x) = x^2 + c$ and $F_\lambda(x) = \lambda x(1 - x)$ and then compare your charts for both of them. Are there any differences? Any similarities? Write a short paragraph about this.

3. Is there a pattern discernible (especially near the period-1 end of your chart)? What is it?

4. Using this pattern, can you predict what the period-1 end of the list at generation 5 will be? Perhaps from the period-1 to the period-A window?

5. *Thought question.* Do you see the "darker" curves running through the orbit diagram? What do you think they represent?

Exercises

For each of the following functions, modify the program in Appendix B to compute the corresponding orbit diagram for the given λ and x intervals. Be sure to use the critical point that lies in the given x-interval to plot the diagram.

1. $S_\lambda(x) = \lambda \sin(x)$, $1 \leq \lambda \leq \pi$, $0 \leq x \leq \pi$.

2. $C_\lambda(x) = \lambda \cos(x)$, $-2.96 \leq \lambda \leq 0$, $-\pi \leq x \leq \pi$.

3. $F_\lambda(x) = \lambda(x - x^3/3)$, $1 \leq \lambda \leq 3\sqrt{3}/2$, $0 \leq x \leq \sqrt{3}$.

The following exercises deal with the family of functions $C_\lambda(x) = \lambda \cos(x)$.

4. Use the computer to sketch the orbit diagram for $1 \leq \lambda \leq 2\pi$, $|x| \leq 2\pi$ using 0 as the critical point. Describe what you see.

5. Explain the dramatic bifurcation that occurs at $\lambda = 2.96 \ldots$.

6. Explain the dramatic bifurcation that occurs at $\lambda = 4.2 \ldots$.

7. What happens if we use the critical point π instead of 0 to plot the orbit diagram? Why?

8. What happens if we let $\lambda > 2\pi$? Discuss other dramatic bifurcations that occur.

The next two exercises deal with the family of functions $\lambda x(1-x)(1-2x)^2$.

9. Use the computer to sketch the orbit diagram for $0 \leq \lambda \leq 16$ and $0 \leq x \leq 1$ using $\frac{1}{2} + \frac{\sqrt{2}}{4} = 0.8535533\ldots$ as the critical point. Describe what you see.

10. Using the graphs of these functions, explain the dramatic bifurcation that occurs at $\lambda = 13.5\ldots$.

The next three exercises deal with the family of "tent maps" given by

$$T_c(x) = \begin{cases} cx & \text{if } 0 \leq x \leq \frac{1}{2} \\ -cx + c & \text{if } \frac{1}{2} < x \leq 1. \end{cases}$$

11. Use the computer to sketch the orbit diagram for T_c for $0 \leq c \leq 2$. Use $1/2$ as the "critical point," since this is the point where T_c is non-differentiable. Describe what you see.

12. Explain the dramatic bifurcation that occurs at $c = 1$.

13. Why do you not see any attracting cycles when $1 \leq c \leq 2$?

All of the following exercises are really essay questions. In each case, provide graphical or numerical evidence for your answer.

14. In the orbit diagram in Figures 8.1 to 8.6, notice that there is a smear near each point where a bifurcation occurs. What causes these smears? It helps to recall the results of Experiment 5.6 here.

15. In Figure 8.11 is plotted the orbit diagram of the logistic function. Instead of using the critical point to plot the diagram, however, we used another initial seed, $x_0 = 0.123\ldots$. Note the change in the diagram. Can you explain what has happened? We will see why this does not occur when we use the critical point in Chapter 12.

16. In Figure 8.5, note the sudden appearance of the period-3 cycle and its window. Explain why this window opens so suddenly.

17. In Figure 8.5, the period-3 window also closes abruptly. Explain why the orbit of 0 suddenly occupies a much larger interval as the period-3 window closes.

x = 1

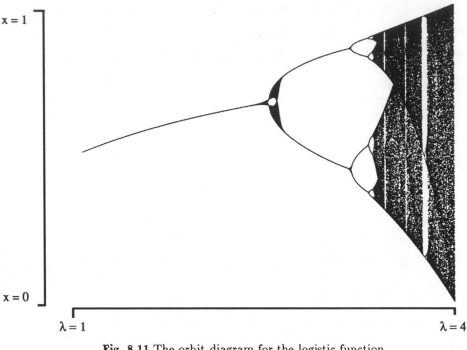

x = 0

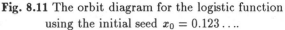

λ = 1   λ = 4

Fig. 8.11 The orbit diagram for the logistic function
using the initial seed $x_0 = 0.123\ldots$.

18. In all of the orbit diagrams, there are some very obvious darker "curves" running through the orbit diagram. Explain what these curves are. Why does the orbit of 0 seem to accumulate on these points? *Hints:* Compare the histograms for various c-values to the slice of the orbit diagram for the corresponding c's. Compute the orbit diagram using only a few points on the orbit of 0 without throwing away any of these points.

CHAPTER 9

Symbolic Dynamics

In this chapter we will introduce one of the most powerful tools for under-standing the chaotic behavior of dynamical systems, symbolic dynamics. We will convert the complicated behavior that we have seen for certain quadra-tic functions to what appears at first to be a completely different dynamical system. However, we will see that the two systems are really the same and that, more importantly, we can understand the new system completely.

This chapter is more theoretical than the preceding chapters. We in-troduce a new level of abstraction involving a "space" of sequences and a mapping on this space that will later serve as a model for the quadratic maps. We will see in the next section that this abstraction is totally justi-fied when we show that iteration of our model mapping can be completely understood. This is in contrast to the quadratic case which is impossible to deal with analytically.

9.1   Itineraries

In this section we will deal exclusively with a quadratic function $Q_c(x) = x^2 + c$ where $c < -2$. Recall that for such a function all of the interesting dynamics occurred on the interval $I = [-p_+, p_+]$. Here p_+ is the fixed point of Q_c given by

$$p_+ = \frac{1}{2}\left(1 + \sqrt{1 - 4c}\right).$$

In this interval there is a subinterval A_1 that consists of all points that leave I under one iteration of Q_c. The set Λ of all points $x \in I$ whose orbits never leave I therefore lies in $I - A_1$. The set $I - A_1$ consists of two closed intervals that we denote by I_0 and I_1 with I_0 to the left of I_1. Remember that both I_0 and I_1 actually depend on c, and for each $c < -2$ these intervals are disjoint and symmetrically located about 0.

Any point in Λ has an orbit that never leaves I and hence remains for all iterations in $I_0 \cup I_1$. So if $x \in \Lambda$, $Q_c^n(x) \in I_0 \cup I_1$ for each n. This allows us to make the following important definition:

Definition. Let $x \in \Lambda$. The *itinerary* of x is the infinite sequence of 0's and 1's given by

$$S(x) = (s_0 s_1 s_2 \ldots)$$

where $s_j = 0$ if $Q_c^j(x) \in I_0$ and $s_j = 1$ if $Q_c^j(x) \in I_1$.

For example, the fixed point p_+ lies in I_1 for all iterations, so p_+ has the itinerary $S(p_+) = (1111\ldots)$. Similarly, the eventually fixed point $-p_+$ has itinerary $S(-p_+) = (01111\ldots)$. Note that any periodic point has an itinerary that is a repeating sequence. A more general itinerary is shown in Figure 9.1.

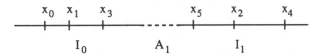

Fig. 9.1 The itinerary of x_0 is $S(x_0) = (001011\ldots)$.

9.2   The Sequence Space

To set up a model system for the dynamics of Q_c on Λ, we need to have a "space" on which the dynamical system takes place. Unlike all of the previous systems we have discussed, this model system will not take place on the real line. Rather, it will take place on a much more abstract space called the *sequence space*.

Definition. The *sequence space* on two symbols is the set

$$\Sigma = \{(s_0 s_1 s_2 \ldots) \mid s_j = 0 \text{ or } 1\}.$$

The set Σ consists of all possible sequences of 0's and 1's. That is, elements of Σ are sequences, not numbers. For example, $(0000\ldots)$, $(010101\ldots)$, $(101010\ldots)$, and $(1111\ldots)$ are all distinct elements of Σ.

We are going to try to work geometrically with the space Σ. That means we have to know what this space looks like. At first, this may sound like an unattainable goal. After all, the elements of Σ are sequences, not numbers that fit conveniently on a line or points that live in the plane or three-dimensional space. However, we will see that with a little metric space theory, we can actually develop a mental image of what Σ looks like.

Let's begin rather naively. Suppose we try to identify Σ as some subset of the plane. Who knows what Σ would look like in the plane? Maybe a triangle or a square or even a banana. Whatever our picture of Σ is, each "point" in Σ must be a sequence. That is, abstractly, we should think of the entire string of digits composing a sequence in Σ as corresponding to one point in this picture. Perhaps Σ is the image sketched in Figure 9.2. If so, then each point in this "space" must be a sequence of 0's and 1's, and two distinct points should be two different sequences, as indicated in the figure.

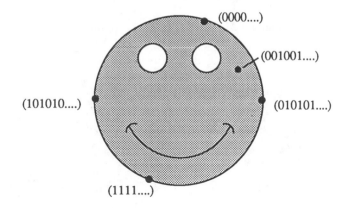

Fig. 9.2 Perhaps the space Σ is a Smiley Face.

We should also be able to recognize certain subsets of Σ. For example, the set

$$M_0 = \{\mathbf{s} \in \Sigma \mid s_0 = 0\}$$

consists of all sequences whose zeroth entry is 0. Its complement is

$$M_1 = \{\mathbf{s} \in \Sigma \mid s_0 = 1\}.$$

Roughly speaking, M_0 and M_1 divide Σ into equal halves, perhaps as depicted in Figure 9.3. Similarly,

$$M_{01} = \{s \in \Sigma \mid s_0 = 0, \, s_1 = 1\}$$

$$M_{00} = \{s \in \Sigma \mid s_0 = 0, \, s_1 = 0\}$$

divide M_0 into equal halves as shown in Figure 9.4.

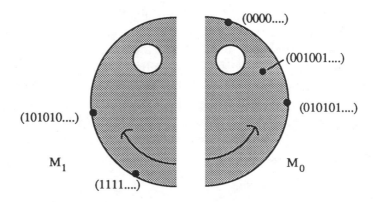

Fig. 9.3 M_0 and M_1 each occupy half of Σ.

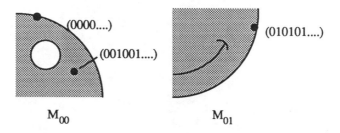

Fig. 9.4 M_{00} and M_{01} each occupy half of M_0.

Of course, given what we know so far (just the definition of Σ), we really have no earthly way of knowing what this set and its subsets look like. What we need is some way of understanding when two points are close together and when they are far apart. That is, we need a way to measure distances between two elements or points in the set. Mathematically, this means we need a *metric* on the set. Now there are many ways to put a metric on a set, just as in real life where there are many ways to measure distances, such as the straight line distance between two points ("as the crow flies") or the distance between the same two points as displayed on a car's odometer. We choose to put the following metric on Σ.

Definition. Let $\mathbf{s} = (s_0 s_1 s_2 \ldots)$ and $\mathbf{t} = (t_0 t_1 t_2 \ldots)$ be two points in Σ. The *distance* between $\mathbf{s}$ and $\mathbf{t}$ is given by

$$d[\mathbf{s}, \mathbf{t}] = \sum_{i=0}^{\infty} \frac{|s_i - t_i|}{2^i}.$$

This may look like a complicated expression, but it is often easy to compute. For example, let $\mathbf{s} = (000\ldots)$, $\mathbf{t} = (111\ldots)$, and $\mathbf{u} = (010101\ldots)$. Then

$$d[\mathbf{s}, \mathbf{t}] = \sum_{i=0}^{\infty} \left(\frac{1}{2}\right)^i = \frac{1}{1 - \frac{1}{2}} = 2$$

$$d[\mathbf{t}, \mathbf{u}] = 1 + \frac{1}{2^2} + \frac{1}{2^4} + \cdots$$
$$= \sum_{i=0}^{\infty} \left(\frac{1}{4}\right)^i = \frac{4}{3}$$

$$d[\mathbf{s}, \mathbf{u}] = \frac{1}{2} + \frac{1}{2^3} + \frac{1}{2^5} + \cdots$$
$$= \frac{1}{2} \sum_{i=0}^{\infty} \left(\frac{1}{4}\right)^i = \frac{2}{3}.$$

Note that the series defining $d[\mathbf{s}, \mathbf{t}]$ always converges. Indeed, s_i and t_i are each either 0 or 1, so $|s_i - t_i| = 0$ or 1. Therefore, this series is dominated by the geometric series $\sum_{i=0}^{\infty} (1/2)^i$, which converges to 2. Thus the farthest apart any two points in Σ may be is 2 units, as in the first example above.

Definition. A function d is called a *metric* on a set X if for any $x, y, z \in X$ the following three properties hold:
  1. $d[x, y] \geq 0$, and $d[x, y] = 0$ if and only if $x = y$.
  2. $d[x, y] = d[y, x]$.
  3. $d[x, z] \leq d[x, y] + d[y, z]$.
The pair X, d is called a *metric space*.

This last property, an important one, is called the *triangle inequality.*
Notice that all three of these properties are familiar properties of the usual
metric or distance function in Euclidean space.

Proposition. *The distance d on Σ given by*

$$d[\mathbf{s}, \mathbf{t}] = \sum_{i=0}^{\infty} \frac{|s_i - t_i|}{2^i}$$

is a metric on Σ.

Proof: Let $\mathbf{s} = (s_0 s_1 s_2 \dots)$, $\mathbf{t} = (t_0 t_1 t_2 \dots)$ and $\mathbf{u} = (u_0 u_1 u_2 \dots)$. Clearly,
$d[\mathbf{s}, \mathbf{t}] \geq 0$ and $d[\mathbf{s}, \mathbf{t}] = 0$ if and only if $\mathbf{s} = \mathbf{t}$. Since $|s_i - t_i| = |t_i - s_i|$, it
follows that $d[\mathbf{s}, \mathbf{t}] = d[\mathbf{t}, \mathbf{s}]$. Finally, for any three real numbers s_i, t_i, u_i, we
have the usual triangle inequality

$$|s_i - t_i| + |t_i - u_i| \geq |s_i - u_i|$$

from which we deduce that

$$d[\mathbf{s}, \mathbf{t}] + d[\mathbf{t}, \mathbf{u}] \geq d[\mathbf{s}, \mathbf{u}].$$

This completes the proof.

The reason we introduced the metric d was to decide when two sequences
in Σ were close together. The next theorem settles this issue.

The Proximity Theorem. *Let $\mathbf{s}, \mathbf{t} \in \Sigma$ and suppose $s_i = t_i$ for $i = 0, 1, \dots, n$. Then $d[\mathbf{s}, \mathbf{t}] \leq 1/2^n$. Conversely, if $d[\mathbf{s}, \mathbf{t}] < 1/2^n$, then $s_i = t_i$ for $i \leq n$.*

Proof. If $s_i = t_i$ for $i \leq n$, then

$$d[\mathbf{s}, \mathbf{t}] = \sum_{i=0}^{n} \frac{|s_i - s_i|}{2^i} + \sum_{i=n+1}^{\infty} \frac{|s_i - t_i|}{2^i}$$

$$= \sum_{i=n+1}^{\infty} \frac{|s_i - t_i|}{2^i} \leq \sum_{i=n+1}^{\infty} \frac{1}{2^i} = \frac{1}{2^n}.$$

On the other hand, if $s_j \neq t_j$ for some $j \leq n$, then we must have

$$d[\mathbf{s}, \mathbf{t}] \geq \frac{1}{2^j} \geq \frac{1}{2^n}.$$

Consequently, if $d[\mathbf{s}, \mathbf{t}] < 1/2^n$, then $s_i = t_i$ for $i \leq n$.

In words, two sequences are close if their first few entries agree. We can guarantee that two sequences are within $1/2^n$ of each other if their first n entries are the same.

This proposition shows that our naive picture of Σ in Figures 9.2 to 9.4 is wrong. For example, any point in M_0 must be at least 1 unit away from any point in M_1, since the initial entries of these sequences disagree. Similarly, any point in M_{00} must be at least $1/2$ unit away from a point in M_{01} (see Figure 9.5). Continuing in this fashion, we see that our picture of Σ cannot contain any planar regions—any such region must eventually be subdivided into infinitely many disjoint pieces. This is, of course, reminiscent of the totally disconnected Cantor set we encountered previously. We will make this analogy precise later.

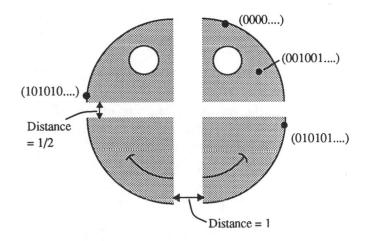

Fig. 9.5 The metric d cuts Σ into disjoint pieces.

9.3   The Shift Map

Now that we have the space that will be the setting for our model dynamical system, it is time to introduce the model mapping itself. This is the *shift map* on Σ.

Definition. The *shift map* $\sigma: \Sigma \to \Sigma$ is defined by

$$\sigma(s_0 s_1 s_2 \ldots) = (s_1 s_2 s_3 \ldots).$$

That is, σ simply drops the first entry of any point in Σ. For example,

$$\sigma(010101\ldots) = (101010\ldots)$$

$$\sigma(101010\ldots) = (010101\ldots)$$

$$\sigma(01111\ldots) = (1111\ldots).$$

Note that it is easy to iterate σ—we simply continue dropping leading entries as we iterate. That is,

$$\sigma^2(s_0 s_1 s_2 \ldots) = (s_2 s_3 s_4 \ldots)$$

$$\sigma^n(s_0 s_1 s_2 \ldots) = (s_n s_{n+1} s_{n+2} \ldots).$$

This makes it easy to find the periodic points of σ. If $\mathbf{s}$ is a repeating sequence of the form

$$\mathbf{s} = (s_0 s_1 \ldots s_{n-1} s_0 s_1 \ldots s_{n-1} \ldots) = \overline{s_0 s_1 \ldots s_{n-1}},$$

then $\sigma^n(\mathbf{s}) = \mathbf{s}$. Conversely, any periodic point of period n for σ must be a repeating sequence. For example, the only fixed points for σ are $(0000\ldots)$ and $(1111\ldots)$. The period 2 points are $(010101\ldots)$ and $(101010\ldots)$, and, of course, σ maps one to the other. There are two 3-cycles for σ given by

$$(\overline{001}) \mapsto (\overline{010}) \mapsto (\overline{100}) \mapsto (\overline{001})$$

$$(\overline{110}) \mapsto (\overline{101}) \mapsto (\overline{011}) \mapsto (\overline{110}).$$

We emphasize how easy it is to find cycles for σ. We may clearly write down all cycles of period n for σ just by writing any block of n 0's and 1's and then repeating it *ad infinitum*. Recall that it would be practically impossible to do this for the quadratic map—this is just one of the reasons why σ is a much simpler mapping to understand.

The shift map is clearly much different from the usual functions encountered in calculus such as polynomials, trigonometric functions, exponentials, and the like. Nevertheless, like its more familiar counterparts, σ is a continuous function. To see why this is true, we cannot use the simplified definition of continuity often given in calculus courses ("you can draw the graph without lifting your pen"). Rather, we have to return to first principles and invoke the theoretical definition of continuity.

Definition. Suppose $F: X \to X$ is a function and X is a set equipped with a metric d. Then F is *continuous* at $x_0 \in X$ if, for any $\epsilon > 0$, there is a $\delta > 0$ such that, if $d[x, x_0] < \delta$ then $d[F(x), F(x_0)] < \epsilon$. We say that F is a *continuous function* if F is continuous at all $x_0 \in X$.

To prove that a function is continuous at x_0, we must therefore be able to accomplish the following: given any $\epsilon > 0$ no matter how small, we must be able to produce a $\delta > 0$ such that, whenever x and x_0 are within δ units of each other, their images are closer together than ϵ.

This definition of continuity is one of those extremely complicated definitions that often drives students mad. So before proving that the shift map is continuous at all points, we'll give a simple example as a warm-up. Let's begin by showing that σ is continuous at the fixed point $(0000\ldots)$. To prove this, we assume that someone gives us an $\epsilon > 0$. Our job is to find a δ that "works." By this we mean that, if $\mathbf{s}$ is a sequence in Σ with $d[\mathbf{s}, (0000\ldots)] < \delta$, then $d[\sigma(\mathbf{s}), (0000\ldots)] < \epsilon$. Now we don't know which sequences in Σ have images that are within an arbitrary ϵ of $(0000\ldots)$, but we do know those that are within $1/2^n$ of $(0000\ldots)$ for each n. And that will be good enough, as long as we choose n so that $1/2^n < \epsilon$. So we must find δ that guarantees that $d[\sigma(s_0 s_1 s_2 \ldots), (0000\ldots)] \leq 1/2^n$. But that's easy! By the Proximity Theorem, the only way this can be true is if $s_i = 0$ for $i = 0, 1, 2, \ldots, n + 1$. Therefore we choose $\delta = 1/2^{n+1}$. If $d[(s_0 s_1 s_2 \ldots), (0000\ldots)] < \delta$, then

$$d[\sigma(s_0 s_1 s_2 \ldots), (0000\ldots)] = d[(s_1 s_2 s_3 \ldots), (0000\ldots)] \leq 1/2^n < \epsilon.$$

Thus we have found a δ that works.

The proof in the general case is not much more difficult.

Theorem. *The function $\sigma: \Sigma \to \Sigma$ is continuous at all points in Σ.*

Proof: Suppose we are given $\epsilon > 0$ and $\mathbf{s} = (s_0 s_1 s_2 \ldots)$. We will show that σ is continuous at $\mathbf{s}$.

Since $\epsilon > 0$, we may pick n such that $1/2^n < \epsilon$. We then choose $\delta = 1/2^{n+1}$. If $\mathbf{t}$ is a point in Σ and $d[\mathbf{t}, \mathbf{s}] < \delta$, then by the Proximity Theorem we must have $s_i = t_i$ for $i = 0, 1, \ldots n + 1$. That is, $\mathbf{t} = (s_0 \ldots s_{n+1} t_{n+2} t_{n+3} \ldots)$.

Now $\sigma(\mathbf{t}) = (s_1 \ldots s_{n+1} t_{n+2} t_{n+3} \ldots)$ has entries that agree with those of $\sigma(\mathbf{s})$ in the first $n + 1$ spots*. Thus, again by the Proximity Theorem,

$$d[\sigma(\mathbf{s}), \sigma(\mathbf{t})] \leq 1/2^n < \epsilon.$$

Therefore, σ is continuous at $\mathbf{s}$. Since $\mathbf{s}$ was arbitrary, σ is a continuous function. This completes the proof.

9.4   Conjugacy

Now that we have our model system, the shift map σ on the sequence space Σ, it is time to relate this dynamical system to Q_c on Λ. We already have a function that relates the two, namely the itinerary function S. Recall that, given a point $x \in \Lambda$, the itinerary of x is a sequence $S(x)$ in Σ. Hence we have a mapping $S: \Lambda \to \Sigma$.

Our first observation is that Λ and Σ, though defined in very different manners, are essentially identical spaces. The way to show that two spaces are identical is to exhibit a *homeomorphism* between them. A homeomorphism is a one-to-one, onto, and continuous function with continuous inverse. If we have a homeomorphism between two sets, these sets are said to be *homeomorphic*. From a qualitative point of view, two sets that are homeomorphic are essentially the same. Not only may we put the points of one set into one-to-one correspondence with those of the other set, but, more importantly, we may do this in a continuous fashion, taking nearby points in one set to nearby points in the other.

Homeomorphism allows for sets to be distorted, as, for example, two finite closed intervals are homeomorphic, even though their lengths differ. But an open interval and a closed interval are never homeomorphic. Rather than interrupt the discussion of symbolic dynamics with a lengthy description of the properties of homeomorphisms at this point, we refer you to Appendix A.1 for a more complete discussion of this topic.

Theorem. (Λ and Σ are the same!) *Suppose $c < -(5 + 2\sqrt{5})/4$. Then $S: \Lambda \to \Sigma$ is a homeomorphism.*

---

*   Technically, these sequences agree in the zeroth through nth spots.

We will conclude this chapter with a proof of this important fact, but before this we turn to another property of S.

Theorem. *If $x \in \Lambda$, then $S \circ Q_c(x) = \sigma \circ S(x)$.*

This result is clear: if $x \in \Lambda$ has itinerary $(s_0 s_1 s_2 \ldots)$, then by definition

$$x \in I_{s_0}$$
$$Q_c(x) \in I_{s_1}$$
$$Q_c^2(x) \in I_{s_2}$$

and so forth. Here, I_{s_j} is either I_0 or I_1, depending upon the digit s_j. Thus

$$Q_c(x) \in I_{s_1}$$
$$Q_c^2(x) \in I_{s_2}$$
$$Q_c^3(x) \in I_{s_3},$$

so that $S(Q_c(x)) = (s_1 s_2 s_3 \ldots)$ which is $\sigma(S(x))$ as claimed.

We may describe this result pictorially as a *commutative diagram:*

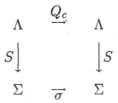

The theorem tells us that if we start with any $x \in \Lambda$ in the upper left-hand corner of this diagram and then follow the arrows in either possible direction, we always end up at the same point in Σ in the lower right-hand corner.

Since S is a homeomorphism, we also have

$$Q_c \circ S^{-1} = S^{-1} \circ S \circ Q_c \circ S^{-1}$$
$$= S^{-1} \circ \sigma \circ S \circ S^{-1}$$
$$= S^{-1} \circ \sigma.$$

Thus, this diagram commutes:

$$
\begin{array}{ccc}
\Lambda & \xrightarrow{Q_c} & \Lambda \\
S^{-1} \uparrow & & \uparrow S^{-1} \\
\Sigma & \xrightarrow{\sigma} & \Sigma
\end{array}
$$

Stephen Smale

After winning the Fields Medal in 1966 for resolving the Poincaré Conjecture, *Stephen Smale* (1930–) turned his attention to dynamical systems. In his famous "horseshoe" example, Smale showed that very simple dynamical systems could behave quite chaotically. Moreover, he introduced the use of symbolic dynamics to help understand this chaotic behavior. The quadratic map in this chapter is a simplified version of the horseshoe example. Smale is currently Professor of Mathematics at the University of California, Berkeley.

We also have

$$S \circ Q_c^n = \sigma \circ S \circ Q_c^{n-1}$$
$$= \sigma^2 \circ S \circ Q_c^{n-2}$$
$$\vdots$$
$$= \sigma^n \circ S.$$

This is an important fact. It says that S converts orbits of Q_c to orbits of σ. Pictorially, we have

$$
\begin{array}{ccccccc}
\Lambda & \xrightarrow{Q_c} & \Lambda & \xrightarrow{Q_c} & \cdots & \xrightarrow{Q_c} & \Lambda \\
S\downarrow & & \downarrow S & & & & \downarrow S \\
\Sigma & \xrightarrow{\sigma} & \Sigma & \xrightarrow{\sigma} & \cdots & \xrightarrow{\sigma} & \Sigma
\end{array}
$$

Similarly, we have

$$Q_c^n \circ S^{-1} = S^{-1} \circ \sigma^n$$

so S^{-1} converts σ-orbits to Q_c-orbits. Therefore, Q_c-orbits are in one-to-one correspondence with σ-orbits. In particular, if $\mathbf{s}$ is a periodic point for σ, then $S^{-1}(\mathbf{s})$ is a periodic point for Q_c with the same period. If we have an eventually periodic point for σ, then S^{-1} gives us an analogous point for Q_c. Indeed, S^{-1} converts any dynamical behavior we observe for σ to corresponding behavior for Q_c. Thus the dynamics of σ on Σ and Q_c on Λ are essentially the same. As we will see in the next chapter, the dynamical behavior of σ is chaotic. Therefore the same is true for Q_c on Λ.

Homeomorphisms like S play an important role in the study of dynamical systems, since they show that two apparently different systems may actually be dynamically equivalent. These maps are called *conjugacies*.

Definition. Let $F: X \to X$ and $G: Y \to Y$ be two functions. We say that F and G are *conjugate* if there is a homeomorphism $h: X \to Y$ such that $h \circ F = G \circ h$. The map h is called a *conjugacy*.

The previous theorems therefore give:

The Conjugacy Theorem. *The shift map on Σ is conjugate to Q_c on Λ when $c < -(5 + 2\sqrt{5})/4$.*

We remark that this result is valid for $-(5 + 2\sqrt{5})/4 < c < -2$, but the arguments needed in this range are more difficult.

Now let's finish the proof that Λ and Σ are homeomorphic. To prove this we need to show that S is one-to-one and onto and that both S and S^{-1} are continuous.

One-to-one: Suppose that $x, y \in \Lambda$ with $x \neq y$. Assume, however, that $S(x) = S(y)$. This means that $Q_c^n(x)$ and $Q_c^n(y)$ always lie in the same subinterval I_0 or I_1. We know that Q_c is one-to-one on each of these intervals and that $|Q_c'(x)| > \mu > 1$ for all $x \in I_0 \cup I_1$ and some μ. Now consider the interval $[x, y]$. For each n, Q_c^n takes this interval in one-to-one fashion onto $[Q_c^n(x), Q_c^n(y)]$. But, as we saw in Chapter 7, the Mean Value Theorem implies that

$$\text{length } [Q_c^n(x), Q_c^n(y)] \geq \mu^n \text{ length } [x, y].$$

Since $\mu^n \to \infty$, we have a contradiction unless $x = y$.

Onto: We first introduce the following notation. Let $J \subset I$ be a closed interval. Let

$$Q_c^{-n}(J) = \{x \in I \mid Q_c^n(x) \in J\}.$$

In particular, $Q_c^{-1}(J)$ denotes the preimage of J. The main observation is that, if $J \subset I$ is a closed interval, then $Q_c^{-1}(J)$ consists of two closed subintervals, one in I_0 and one in I_1. (see Figure 9.6; also consult Appendix A.3).

To find $x \in \Lambda$ with $S(x) = \mathbf{s}$, we define

$$I_{s_0 s_1 \ldots s_n} = \{x \in I \mid x \in I_{s_0}, Q_c(x) \in I_{s_1}, \ldots, Q_c^n(x) \in I_{s_n}\}.$$

Since $s_j = 0$ or 1 for each j, the set I_{s_j} is really one of I_0 or I_1 depending on the digit s_j. Using the notation above we may write

$$I_{s_0 s_1 \ldots s_n} = I_{s_0} \cap Q_c^{-1}(I_{s_1}) \cap \ldots \cap Q_c^{-n}(I_{s_n}).$$

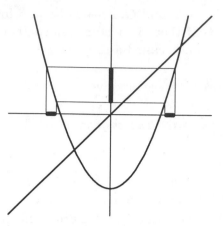

Fig. 9.6 The preimage of a closed interval J is a
pair of closed intervals, one in I_0 and one in I_1.

For any pair of intervals A and B, we have

$$Q_c^{-1}(A \cap B) = Q_c^{-1}(A) \cap Q_c^{-1}(B).$$

Thus we may also write

$$I_{s_0 s_1 \ldots s_n} = I_{s_0} \cap Q_c^{-1}\left(I_{s_1} \cap \ldots \cap Q_c^{-(n-1)}(I_{s_n})\right)$$
$$= I_{s_0} \cap Q_c^{-1}(I_{s_1 \ldots s_n}).$$

We claim that the $I_{s_0 s_1 \ldots s_n}$ are closed intervals that are nested. Clearly I_{s_0}
is a closed interval. By induction we assume that $I_{s_1 \ldots s_n}$ is a closed interval.
Then, by the observation above, $Q_c^{-1}(I_{s_1 \ldots s_n})$ consists of a pair of intervals,
one in I_0 and one in I_1. In either event, $I_{s_0} \cap Q_c^{-1}(I_{s_1 \ldots s_n}) = I_{s_0 s_1 \ldots s_n}$ is a
single closed interval.

These intervals are nested because

$$I_{s_0 \ldots s_n} = I_{s_0 \ldots s_{n-1}} \cap Q_c^{-n}(I_{s_n}) \subset I_{s_0 \ldots s_{n-1}}.$$

Therefore we conclude that

$$\bigcap_{n \geq 0} I_{s_0 s_1 \ldots s_n}$$

is nonempty since a nested intersection of closed intervals is always nonempty.
See Appendix A.3 for a discussion of this fact. Note that if $x \in \cap_{n \geq 0} I_{s_0 s_1 \ldots s_n}$,
then $x \in I_{s_0}$, $Q_c(x) \in I_{s_1}$, and so forth. Hence $S(x) = (s_0 s_1 \ldots)$. This proves
that S is onto.

Incidentally, note that $\cap_{n \geq 0} I_{s_0 s_1 \ldots s_n}$ consists of a unique point. This follows immediately from the fact that S is one-to-one. In particular, we have that diam $I_{s_0 s_1 \ldots s_n} \to 0$ as $n \to \infty$.

Continuity: To prove that S is continuous, we again invoke the theoretical definition of continuity given above. Let $x \in \Lambda$ and suppose that $S(x) = s_0 s_1 s_2 \ldots$. We will show that S is continuous at x. Let $\epsilon > 0$. Then pick n so that $1/2^n < \epsilon$. Consider the closed subintervals $I_{t_0 t_1 \ldots t_n}$ defined above for all possible combinations $t_0 t_1 \ldots t_n$. These subintervals are all disjoint, and Λ is contained in their union. There are 2^{n+1} such subintervals, and $I_{s_0 s_1 \ldots s_n}$ is one of them. Hence we may choose δ such that $|x - y| < \delta$ and $y \in \Lambda$ implies that $y \in I_{s_0 s_1 \ldots s_n}$. To do this we simply choose δ so small that the interval of length 2δ centered at x overlaps none of the $I_{t_0 t_1 \ldots t_n}$ with the exception of $I_{s_0 s_1 \ldots s_n}$. Therefore, $S(y)$ agrees with $S(x)$ in the first $n + 1$ terms. Hence, by the Proximity Theorem, we have

$$d[S(x), S(y)] \leq \frac{1}{2^n} < \epsilon.$$

This proves the continuity of S. It is easy to check that S^{-1} is also continuous. Thus, S is a homeomorphism.

Exercises

1. List all cycles of prime period 4 for the shift map.

Compute $d[\mathbf{s}, \mathbf{t}]$ where:

2. $\mathbf{s} = (\overline{100})$, $\mathbf{t} = (\overline{001})$.

3. $\mathbf{s} = (\overline{100})$, $\mathbf{t} = (\overline{010})$.

4. $\mathbf{s} = (\overline{1011})$, $\mathbf{t} = (01\overline{01})$.

5. Find all points in Σ whose distance from $(000 \ldots)$ is exactly $1/2$.

6. Give an example of a sequence midway between $(000 \ldots)$ and $(111 \ldots)$. Give a second such example. Are there any other such points? Why or why not?

7. Let $M_{01} = \{\mathbf{s} \in \Sigma \,|\, s_0 = 0, s_1 = 1\}$ and $M_{101} = \{\mathbf{s} \in \Sigma \,|\, s_0 = 1, s_1 = 0, s_2 = 1\}$. What is the minimum distance between a point in M_{01} and a point in M_{101}? Give an example of two sequences that are this close to each other.

8. What is the maximum distance between a point in M_{01} and a point in M_{101}? Give an example of two sequences that are this far apart.

9. Let S be the itinerary map as defined in this chapter. Prove that if $\mathbf{s} \in \Sigma$ is periodic under the shift map, then $S^{-1}(\mathbf{s})$ is a periodic point for Q_c in Λ with the same period. What happens if $\mathbf{s}$ is eventually periodic?

The N–Shift:

The following seven exercises deal with the analog of the shift map and sequence space for sequences that have more than two possible entries, the space of sequences of N symbols.

10. Let Σ_N denote the space of sequences whose entries are the positive integers $0, 1, \ldots, N - 1$, and let σ_N be the shift map on Σ_N. For $\mathbf{s}, \mathbf{t} \in \Sigma_N$, let

$$d_N[\mathbf{s}, \mathbf{t}] = \sum_{i=0}^{\infty} \frac{|s_i - t_i|}{N^i}.$$

Prove that d_N is a metric on Σ_N.

11. What is the maximal distance between two sequences in Σ_N?

12. How many fixed points does σ_N have? How many 2-cycles? How many cycles of prime period 2?

13. How many points in Σ_N are fixed by σ_N^k?

14. Prove that $\sigma_N : \Sigma_N \to \Sigma_N$ is continuous.

15. Now define

$$d_\delta[\mathbf{s}, \mathbf{t}] = \sum_{k=0}^{\infty} \frac{\delta_k(\mathbf{s}, \mathbf{t})}{N^k}$$

where $\delta_k(\mathbf{s}, \mathbf{t}) = 0$ if $s_k = t_k$ and $\delta_k(\mathbf{s}, \mathbf{t}) = 1$ if $s_k \neq t_k$. Prove that d_δ is also a metric on Σ_N.

16. What is the maximum distance between two points in Σ_N when the metric d_δ is used?

17. Recall the function

$$T(x) = \begin{cases} 3x & \text{if } x \leq 1/2 \\ 3 - 3x & \text{if } x > 1/2 \end{cases}$$

that was discussed in the exercises at the end of Chapter 7. There we proved that $\Gamma = \{x \in [0,1] \mid T^n(x) \in [0,1] \text{ for all } n\}$ was the Cantor middle-thirds set. Now define an itinerary function $S : \Gamma \to \Sigma$, where Σ is the space of sequences of 0's and 1's. Prove that S is a homeomorphism.

Remarks:

1. The above exercise shows that the set Λ for the quadratic map and the Cantor middle-thirds set are actually homeomorphic, for they are both homeomorphic to Σ.

2. We now have two infinite sequences attached to each point x in the Cantor set. One is the ternary expansion of x and the other is the itinerary of x. It is tempting to seek a relationship between these two sequences. While there is a relationship between them, it is by no means obvious. For example, the point 1 has ternary expansion $0.222\ldots$ but $S(1) = (1000\ldots)$.

18. Each of the following defines a function on the space of sequences Σ. In each case, decide if the given function is continuous. If so, prove it. If not, explain why.

a. $F(s_0 s_1 s_2 \ldots) = (0 s_0 s_1 s_2 \ldots)$
b. $G(s_0 s_1 s_2 \ldots) = (0 s_0 0 s_1 0 s_2 \ldots)$
c. $H(s_0 s_1 s_2 \ldots) = (s_1 s_0 s_3 s_2 s_5 s_4 \ldots)$
d. $J(s_0 s_1 s_2 \ldots) = (\hat{s}_0 \hat{s}_1 \hat{s}_2 \ldots)$ where $\hat{s}_j = 1$ if $s_j = 0$ and $\hat{s}_j = 0$ if $s_j = 1$.
e. $K(s_0 s_1 s_2 \ldots) = ((1 - s_0)(1 - s_1)(1 - s_2)\ldots)$
f. $L(s_0 s_1 s_2 \ldots) = (s_0 s_2 s_4 s_6 \ldots)$
g. $M(s_0 s_1 s_2 \ldots) = (s_1 s_{10} s_{100} s_{1000} \ldots)$
h. $N(s_0 s_1 s_2 \ldots) = (t_0 t_1 t_2 \ldots)$ where $t_j = s_0 + s_1 + \cdots + s_j \mod 2$. That is, $t_j = 0$ if $s_0 + \cdots + s_j$ is even, and $t_j = 1$ if $s_0 + \cdots + s_j$ is odd.
i. $P(s_0 s_1 s_2 \ldots) = (t_0 t_1 t_2 \ldots)$ where $t_j = \lim_{n \to \infty} s_n$ if this limit exists. Otherwise, $t_j = s_j$.

19. Define a different distance function d' on Σ by $d'[\mathbf{s}, \mathbf{t}] = 1/(k+1)$ where k is the least index for which $s_k \neq t_k$ and $d'[\mathbf{s}, \mathbf{s}] = 0$. Is d' a metric?

CHAPTER 10

Chaos

In this chapter we will introduce the notion of chaos. We will show that there are many dynamical systems that are chaotic but that nevertheless can be completely understood. We will describe how this can happen first for the shift map and then later in other important examples, including the quadratic family.

10.1   Three Properties of a Chaotic System

There are many possible definitions of chaos. In fact, there is no general agreement within the scientific community as to what constitutes a chaotic dynamical system. However, this will not deter us from offering one possible definition. This definition has the advantage that it may be readily verified in a number of different and important examples. However, you should be forewarned that there are many other possible ways to capture the essence of chaos.

To describe chaos, we need one preliminary notion from topology, that of a *dense set*.

Definition. Suppose X is a set and Y is a subset of X. We say that Y is *dense* in X if, for any point $x \in X$, there is a point y in the subset Y arbitrarily close to x.*

---

*   To make the notion of "closeness" precise, we must have a metric or distance function on the set X.

Equivalently, Y is dense in X if for any $x \in X$ we can find a sequence of points $\{y_n\} \in Y$ that converges to x. For example, the subset of rational numbers is dense in the set of real numbers. So is the subset consisting of all irrational numbers. However, the integers are far from being dense in the reals. Finally, the open interval (a, b) is dense in the closed interval $[a, b]$.

To prove that a subset $Y \subset X$ is dense in X, we must exhibit a sequence of points in Y that converges to an arbitrary point in X. For example, to prove that the rational numbers are dense in $\mathbf{R}$, we must find a sequence of rationals converging to any irrational. For instance, if the irrational is $\sqrt{2}$, a sequence of rationals converging to this number is

$$1, 1.4, 1.41, 1.414, \ldots.$$

In the general case, we begin by selecting an arbitrary real number x. If x is rational, then we are done, so we assume that x is irrational. This means that x has an infinite decimal expansion of the form

$$x = a_n \ldots a_0.b_1 b_2 b_3 \ldots$$

where the a_j and b_j are digits ranging from 0 to 9. Now, for $j = 1, 2, 3, \ldots$, set

$$x_j = a_n \ldots a_0.b_1 \ldots b_j.$$

Since x_j has a finite decimal expansion, x_j is a rational number. Clearly, $x_j \rightarrow x$ as $j \rightarrow \infty$. So we have found a sequence of rational numbers that converges to x. This proves density of the rationals.

It is tempting to think of a dense subset as being a relatively large subset of a given set. This is true in the sense that there are points in this subset arbitrarily close to any given point in the larger set. However, a dense set may sometimes be small in the sense that it contains only countably many points, as the example of the rationals in the reals shows.

Here is another way in which the set of rationals, though dense in the reals, is small. Let R be the subset of the interval $[0, 1]$ that consists of all of the rationals in $[0, 1]$. We may list all of the elements of R. One such listing is

$$0, 1, \frac{1}{2}, \frac{1}{3}, \frac{2}{3}, \frac{1}{4}, \frac{3}{4}, \frac{1}{5}, \frac{2}{5}, \frac{3}{5}, \frac{4}{5}, \frac{1}{6}, \ldots.$$

Now let ϵ be small. Consider the interval of length ϵ^n about the nth element in the above list. The union of all of these intervals is clearly an open set. Its intersection with the interval $[0, 1]$ is dense since it contains all of the rationals. However, the total length of this set is small. Indeed, the total length is given by

$$\sum_{n=1}^{\infty} \epsilon^n = \frac{\epsilon}{1-\epsilon},$$

which is small when ϵ is very small. For example, the total length of this set is $1/99$ when $\epsilon = 0.01$.

Now let's return to investigate the dynamics of the shift map σ on the sequence space Σ. Our first observation about this map is that the subset of Σ that consists of all periodic points in Σ is a dense subset. To see why this is true, we must show that, given any point $\mathbf{s} = (s_0 s_1 s_2 \ldots)$ in Σ, we can find a periodic point arbitrarily close by. So suppose we are given an $\epsilon > 0$. How do we find a periodic point within ϵ units of $\mathbf{s}$? Let's choose an integer n so that $1/2^n < \epsilon$. We may now write down an explicit periodic point within $1/2^n$ units of $\mathbf{s}$. Let $\mathbf{t}_n = (s_0 s_1 \ldots s_n \overline{s_0 s_1 \ldots s_n})$. The first $n+1$ entries of $\mathbf{s}$ and $\mathbf{t}_n$ are the same. By the Proximity Theorem this means that

$$d[\mathbf{s}, \mathbf{t}_n] \leq \frac{1}{2^n} < \epsilon.$$

But $\mathbf{t}_n$ is a repeating sequence and so it is a periodic point of period $n+1$ for σ. Since ϵ and $\mathbf{s}$ were arbitrary, we have succeeded in finding a periodic point arbitrarily close to any point in Σ. Note that the sequence of sequences $\{\mathbf{t}_n\}$ converges to $\mathbf{s}$ as $n \to \infty$.

Note again the power of symbolic dynamics. Unlike maps like the quadratic function, we can explicitly exhibit periodic points of any period for σ. Moreover, we can show that these points accumulate or come arbitrarily close to any given point in Σ.

A second and even more interesting property of σ is that there is a point whose orbit is dense in Σ. That is to say, we can find an orbit which comes arbitrarily close to any point whatsoever in Σ. Clearly, this kind of orbit is far from periodic or eventually periodic. As above, we can write down such an orbit explicitly for σ. Consider the point

$$\hat{\mathbf{s}} = (\ \underbrace{0\ 1}_{1blocks}\ \ \underbrace{00\ 01\ 10\ 11}_{2blocks}\ \ \underbrace{000\ 001\ldots}_{3blocks}\ \ \underbrace{\ldots}_{4blocks}\ ).$$

In words, $\hat{\mathbf{s}}$ is the sequence which consists of all possible blocks of 0's and 1's of length 1, followed by all such blocks of length 2, then length 3, and so forth.

The point $\hat{\mathbf{s}}$ has an orbit that forms a dense subset of Σ. To see this, we again choose an arbitrary $\mathbf{s} = (s_0 s_1 s_2 \ldots) \in \Sigma$ and an $\epsilon > 0$. Again choose n

so that $1/2^n < \epsilon$. Now we show that the orbit of $\hat{\mathbf{s}}$ comes within $1/2^n$ units of $\mathbf{s}$. Far to the right in the expression for $\hat{\mathbf{s}}$, there is a block of length $n + 1$ that consists of the digits $s_0 s_1 \ldots s_n$. Suppose the entry s_0 is at the kth place in the sequence. Now apply the shift map k times to $\hat{\mathbf{s}}$. Then the first $n + 1$ entries of $\sigma^k(\hat{\mathbf{s}})$ are precisely $s_0 s_1 \ldots s_n$. So by the Proximity Theorem,

$$d[\sigma^k(\hat{\mathbf{s}}), \mathbf{s}] \leq \frac{1}{2^n} < \epsilon.$$

There is a dynamical notion that is intimately related to the property of having a dense orbit. This is the concept of *transitivity*.

Definition. A dynamical system is *transitive* if for any pair of points x and y and any $\epsilon > 0$ there is a third point z within ϵ of x whose orbit comes within ϵ of y.

In other words, a transitive dynamical system has the property that, given any two points, we can find an orbit that comes arbitrarily close to both. Clearly, a dynamical system that has a dense orbit is transitive, for the dense orbit comes arbitrarily close to all points. The fact is that the converse is also true—a transitive dynamical system has a dense orbit.* However, we will not prove this fact here since it uses an advanced result from real analysis known as the Baire Category Theorem.

A third property exhibited by the shift map is *sensitive dependence on initial conditions*, or *sensitivity* for short.

Definition. A dynamical system F *depends sensitively on initial conditions* if there is a $\beta > 0$ such that for any x and any $\epsilon > 0$ there is a y within ϵ of x and a k such that the distance between $F^k(x)$ and $F^k(y)$ is at least β.

In this definition it is important to understand the order of the quantifiers. The definition says that, no matter which x we begin with and no matter how small a region we choose about x, we can always find a y in this region whose orbit eventually separates from that of x by at least β. Moreover, the distance β is independent of x. As a consequence, for each

---

* This is true for dynamical systems on the kinds of spaces we are considering, like the real line or Cantor sets. But it need not be true on certain "pathological" kinds of spaces.

Edward N. Lorenz

Edward N. Lorenz (1917–) began his career as a mathematics graduate student at Harvard, but after World War II, turned his attention to meteorology. In 1961, using a primitive computer by today's standards, Lorenz attempted to solve a much-simplified model for weather prediction. His model seemed to simulate real weather patterns quite well, but it also illustrated something much more important: when Lorenz changed the initial conditions in the model slightly, the resulting weather patterns changed completely after a very short time. Lorenz had discovered the fact that very simple differential equations could possess sensitive dependence on initial conditions. Moreover, the Lorenz model has been shown to be a *strange attractor*, an extremely complicated geometric object on which the study of the solutions of the differential equation reduces to the study of the chaotic behavior of an iterated function. Lorenz is currently Professor Emeritus of Meteorology at MIT.

---

x, there are points arbitrarily nearby whose orbits are eventually "far" from that of x.

Remarks:

1. The definition of sensitivity does *not* require that the orbit of y remain far from x for all iterations. We only need one point on the orbit to be far from the corresponding iterate of x.

2. There are other possible definitions of sensitive dependence. For example, one common definition requires that certain nearby orbits diverge exponentially. That is, it is sometimes required that the distance between $F^k(x)$ and $F^k(y)$ grow like $C\mu^k$ for some $\mu > 1$ and $C > 0$.

3. The concept of sensitive dependence on initial conditions is a very important notion in the study of applications of dynamical systems. If a particular system possesses sensitive dependence, then for all practical purposes, the dynamics of this system defy numerical computation. Small errors in computation that are introduced by round-off may throw us off the intended orbit. Then these errors may become magnified upon iteration. Also, as always happens in real-life systems, we can never know the exact initial point of our system no matter how many digits of accuracy we use. As a consequence, we may be looking at an orbit that eventually diverges from the true orbit we

seek. Therefore, the results of numerical computation of an orbit, no matter how accurate, may bear no resemblance whatsoever to the real orbit.

Example. The function $C(x) = \cos x$ possesses no sensitivity to initial conditions whatsoever. Indeed, as we saw in Fig. 4.3, all orbits of C tend to the attracting fixed point at $0.73908\ldots$. On the other hand, $F(x) = \sqrt{x}$ has sensitive dependence at 0. See Fig. 4.2. Although 0 is a fixed point, any nearby point has orbit that tends to the attracting fixed point at 1, hence "far away." On the other hand, there is no sensitive dependence in the interval $0 < x < \infty$.

To see that the shift map depends sensitively on initial conditions, we select $\beta = 1$. For any $\mathbf{s} \in \Sigma$ and $\epsilon > 0$ we again choose n so that $1/2^n < \epsilon$. Suppose $\mathbf{t} \in \Sigma$ satisfies $d[\mathbf{s}, \mathbf{t}] < 1/2^n$ but $\mathbf{t} \neq \mathbf{s}$. Then we know that $t_i = s_i$ for $i = 0, \ldots, n$. However, since $\mathbf{t} \neq \mathbf{s}$ there is $k > n$ such that $s_k \neq t_k$. So $|s_k - t_k| = 1$.

Now consider the sequences $\sigma^k(\mathbf{s})$ and $\sigma^k(\mathbf{t})$. The initial entries of each of these sequences are different, so we have

$$d[\sigma^k(\mathbf{s}), \sigma^k(\mathbf{t})] \geq \frac{|s_k - t_k|}{2^0} + \sum_{i=1}^{\infty} \frac{0}{2^i} = 1.$$

This proves sensitivity for the shift.

Note that we have actually proved a lot more for the shift. We have actually shown that for any $\mathbf{s} \in \Sigma$, *all* other points have orbits that eventually separate by at least 1 unit from the orbit of $\mathbf{s}$.

These three properties are the basic ingredients of a chaotic system:

Definition. A dynamical system F is *chaotic* if:
 1. Periodic points for F are dense.
 2. F is transitive.
 3. F depends sensitively on initial conditions.

So we have proved

Theorem. *The shift map* $\sigma: \Sigma \to \Sigma$ *is a chaotic dynamical system.*

Remark. In fact, it is known that a system that has a dense set of periodic points and is transitive also depends sensitively on initial conditions, so condition three above follows from the first two.[*]

---

[*]   J. Banks, et. al. "On Devaney's Definition of Chaos" *Amer. Math. Monthly* **99** (1992), 332-334.

As we saw in the previous chapter, the shift map on Σ and the quadratic map Q_c on Λ are conjugate and therefore dynamically equivalent. It is natural to ask if the analog of this theorem therefore holds for the quadratic map. Indeed it does, but to show this we first have to make one observation about dense subsets.

The Density Proposition. *Suppose $F: X \to Y$ is a continuous map that is onto and suppose also that $D \subset X$ is a dense subset. Then $F(D)$ is dense in Y.*

Proof: Suppose $y_0 \in Y$ and $\epsilon > 0$. We must produce a point $z \in F(D)$ within ϵ of y_0. Consider instead a preimage $x_0 \in X$ of y_0, that is, $F(x_0) = y_0$. We can find such an x_0 since F is onto. Since F is also continuous, there is $\delta > 0$ such that if x is within δ of x_0, then $F(x)$ is within ϵ of $F(x_0)$. Since D is dense in X, we may choose $\hat{x} \in D$ within δ of x_0. Therefore $F(\hat{x})$ lies in $F(D)$ within ϵ of y_0. So we set $z = F(\hat{x})$. Since both ϵ and y_0 were arbitrary, this completes the proof.

Theorem. *Suppose $c < -(5 + 2\sqrt{5})/4$. Then the quadratic map $Q_c(x) = x^2 + c$ is chaotic on the set Λ.*

Proof: Since the itinerary map $S: \Lambda \to \Sigma$ is a conjugacy, it follows that $S^{-1}: \Sigma \to \Lambda$ is a homeomorphism. Therefore the Density Proposition guarantees that the set of periodic points for Q_c is dense in Λ, since S^{-1} carries periodic points for σ to periodic points for Q_c. Also, if $\hat{s}$ lies on a dense orbit for σ, then the Density Proposition also guarantees that the orbit of $S^{-1}(\hat{s})$ lies on a dense orbit for Q_c. So to prove that Q_c is chaotic, all we need to do is exhibit sensitive dependence.

To accomplish this we need to find a $\beta > 0$ that "works." Recall from Figure 7.5 that Λ is contained in the union of two closed intervals I_0 and I_1 which are disjoint. Choose β to be smaller than the minimum distance between these two intervals. We now claim that *any* two Q_c-orbits eventually separate by at least β. To see this, let $x, y \in \Lambda$ with $x \neq y$. Since S is a homeomorphism, $S(x) \neq S(y)$ as well. As a consequence, these two sequences differ at some entry, say the kth. This means that $F^k(x)$ and $F^k(y)$ each lie in a different I_j. Hence the distance between $F^k(x)$ and $F^k(y)$ is at least β. Therefore any orbit close to x eventually separates from the orbit of x by at least β and we are done.

To summarize, a chaotic map possesses three ingredients: unpredictability, indecomposability, and an element of regularity. A chaotic system is unpredictable because of the sensitive dependence on initial conditions. It cannot be broken down or decomposed into two subsystems that do not interact under F because of transitivity. And, in the midst of this complicated behavior, we nevertheless have an element of regularity, namely the periodic points that are dense.

10.2 Other Chaotic Systems

In the previous section, we showed that the quadratic map Q_c was chaotic on the set Λ as long as c was sufficiently negative. In one sense, this result is unsatisfying because the set on which Q_c is chaotic is "small"—as we will soon see, Λ is a Cantor set. Indeed, it is difficult to see this chaotic behavior with a computer, as we observed in Experiment 3.6. On the other hand, the quadratic function $Q_{-2}(x) = x^2 - 2$ seems to be chaotic on the entire interval $[-2, 2]$, as we saw numerically in Figure 3.3. Our goal in this section is to verify this fact.

Instead of dealing directly with $x^2 - 2$, we will again take a back-door approach and consider a simpler system that turns out to be equivalent to this map. Consider the function $V(x) = 2|x| - 2$. The graph of V is displayed in Figure 10.1. Note that this graph takes the interval $[-2, 2]$ to itself, exactly as $x^2 - 2$ does. Graphical analysis shows that if $|x| > 2$, then the orbit of x under V tends to infinity, again exactly as happened for $x^2 - 2$.

To compute higher iterates of V, we first make use of the definition of absolute value to write

$$
\begin{aligned}
V^2(x) &= 2\,|2|x| - 2| - 2 \\
&= |4|x| - 4| - 2 \\
&= \begin{cases} 4|x| - 6 & \text{if } 4|x| - 4 \geq 0 \iff |x| \geq 1 \\ -4|x| + 2 & \text{if } 4|x| - 4 \leq 0 \iff |x| \leq 1. \end{cases}
\end{aligned}
$$

In turn, this may be further decomposed to

$$
V^2(x) = \begin{cases} 4x - 6 & \text{if } x \geq 1 \\ -4x + 2 & \text{if } 0 \leq x \leq 1 \\ 4x + 2 & \text{if } -1 \leq x \leq 0 \\ -4x - 6 & \text{if } x \leq -1. \end{cases}
$$

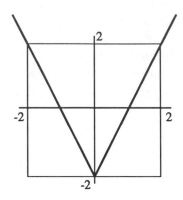

Fig. 10.1 The graph of $V(x) = 2|x| - 2$.

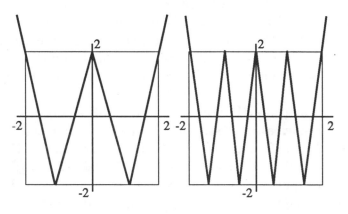

Fig. 10.2 The graphs of V^2 and V^3.

Figure 10.2 shows the graphs of V^2 and V^3. Note that the graph of V^2 consists of four linear pieces, each with slope ± 4, and that the graph of V^3 consists of eight pieces, each with slope ± 8. In general, the graph of V^n consists of 2^n pieces, each of which is a straight line with slope $\pm 2^n$. Each of these linear portions of the graph is defined on an interval of length $1/2^{n-2}$.

This fact shows immediately that V is chaotic on $[-2, 2]$. To see this, we consider an open subinterval J in $[-2, 2]$. From the above observation, we may always find a subinterval of J of length $1/2^{n-2}$ on which the graph of V^n stretches from -2 to 2 (see Fig. 10.3). In particular, V^n has a fixed point in J, so this proves that periodic points are dense in $[-2, 2]$. Also, the image of J covers the entire interval $[-2, 2]$, so V is transitive. Finally, for any $x \in J$, there is a $y \in J$ such that $|V^n(x) - V^n(y)| \geq 2$. Thus we may choose $\beta = 2$ and we have sensitive dependence on initial conditions.

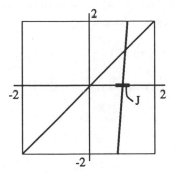

Fig. 10.3 The graph of V^n stretches J over the entire interval $[-2, 2]$.

Now we will use this fact to prove that Q_{-2} is also chaotic. Consider the function $C(x) = -2\cos(\pi x/2)$. This function maps the interval $[-2, 2]$ onto itself as shown in Figure 10.4. Every point in $[-2, 2]$ has exactly two preimages in $[-2, 2]$ with the exception of -2 which has only one.

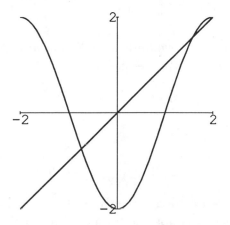

Fig. 10.4 $C(x) = -2\cos(\pi x/2)$ takes $[-2, 2]$ onto itself.

Now suppose we apply C to both x and $V(x)$. The result is the diagram:

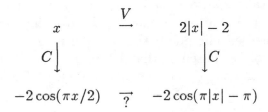

We ask what is the mapping F that completes this diagram: what is the function that accomplishes the following?

$$-2\cos(\pi x/2) \mapsto -2\cos(\pi|x| - \pi).$$

With the help of a little trigonometry, we find

$$
\begin{aligned}
-2\cos(\pi|x| - \pi) &= 2\cos(\pi|x|) \\
&= 2\cos(\pi x) \\
&= 2\cos\left(2 \cdot \frac{\pi x}{2}\right) \\
&= \left(2\cos\left(\frac{\pi x}{2}\right)\right)^2 - 2 \\
&= \left(-2\cos\left(\frac{\pi x}{2}\right)\right)^2 - 2.
\end{aligned}
$$

That is, F is the function that satisfies

$$-2\cos(\pi x/2) \xrightarrow{F} \left(-2\cos\left(\frac{\pi x}{2}\right)\right)^2 - 2,$$

or, letting $u = -2\cos(\pi x/2)$, we find that $F(u) = u^2 - 2 = Q_{-2}(u)$. That is, we have the following commutative diagram:

$$
\begin{array}{ccc}
 & V & \\
[-2,2] & \longrightarrow & [-2,2] \\
C \downarrow & & \downarrow C \\
[-2,2] & \xrightarrow[Q_{-2}]{} & [-2,2]
\end{array}
$$

It appears that C is a conjugacy between V and Q_{-2}. However, C is not a homeomorphism since C is not one-to-one. Nevertheless, the commutative diagram shows that C carries orbits of V to orbits of Q_{-2}. Since C is at most two-to-one, it follows that C takes cycles to cycles. However, it is not necessarily true that C preserves the period of cycles. For example, C could conceivably take a 2-cycle for V and map it to a fixed point for Q_{-2}. Nevertheless, since C is both continuous and onto, the Density Proposition shows that Q_{-2} has periodic points that are dense as well as a dense orbit. Finally, since n may be chosen so that V^n maps arbitrarily small intervals onto all of $[-2, 2]$, the same must be true for Q_{-2}. This proves sensitive dependence and we have:

Theorem. *The function $Q_{-2}(x) = x^2 - 2$ is chaotic on $[-2, 2]$.*

The map C that converts orbits of V to orbits of Q_{-2} has a name. Such a map is called a semiconjugacy. To be precise, we define:

Definition. Suppose $F: X \to X$ and $G: Y \to Y$ are two dynamical systems. A mapping $h: X \to Y$ is called a *semiconjugacy* if h is continuous, onto, at most n-to-one, and satisfies

$$h \circ F = G \circ h.$$

As a final example of a chaotic dynamical system, we consider now a function that will play a major role later when we consider dynamical systems in the complex plane. Let S^1 denote the unit circle in the plane.* That is,

$$S^1 = \{(x, y) \in \mathbf{R}^2 \,|\, x^2 + y^2 = 1\}.$$

We describe a point on S^1 by giving its polar angle θ in radians. Note that θ is only defined modulo 2π.

Let $D: S^1 \to S^1$ be given by $D(\theta) = 2\theta$. We call D the *doubling map* on the circle. Note that if L is an arc on the circle, then $D(L)$ is an arc that is twice as long (unless the arclength of L exceeds π, in which case $D(L)$ covers the entire circle). So one iteration of D behaves in a fashion similar to V above.

Theorem. *The doubling map D is chaotic on the unit circle.*

Proof: To prove this we will again make use of a semiconjugacy. Define the function $B: S^1 \to [-2, 2]$ given by $B(\theta) = 2\cos(\theta)$. Since $\cos(\theta)$ is the x-coordinate of the point θ on S^1, the map B is given geometrically by projecting points vertically from the circle to the x-axis, and then stretching by a factor of 2 (see Fig. 10.5). Note that B is two-to-one except at the points π and 0 on S^1.

Consider the diagram

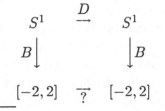

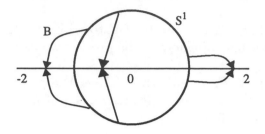

Fig. 10.5 The map B projects S^1 to the x-axis.

As before we ask which function completes the diagram. We have

$$B \circ D(\theta) = 2\cos(2\theta),$$

so we must find the function that takes

$$2\cos(\theta) \mapsto 2\cos(2\theta).$$

However, we may write

$$2\cos(2\theta) = 2\left(2\cos^2(\theta) - 1\right) = (2\cos(\theta))^2 - 2.$$

So the required function is our friend the quadratic function $Q_{-2}(x) = x^2 - 2$. Thus D and Q_{-2} are semiconjugate. It is not difficult to mimic the arguments given above to complete the proof.

10.3   Manifestations of Chaos

There are a number of different computer experiments that give an indication of chaotic behavior. For example, a histogram or density plot can often indicate the presence of orbits that visit virtually every region of the interval. In Figure 10.6 we display the histograms for several functions, including $F_4(x) = 4x(1 - x)$ and $G(x) = 4x^3 - 3x$. These histograms are computed using 30,000 iterations of a randomly chosen initial seed in each case. Note the similarity between these histograms and that of the quadratic function Q_{-2} displayed in Figure 3.3. We ask you to verify that each of these maps is indeed chaotic in the exercises following this chapter. We have also displayed the histograms for $Q_{-1.8}$ and $Q_{-1.6}$ in Figure 10.6. These

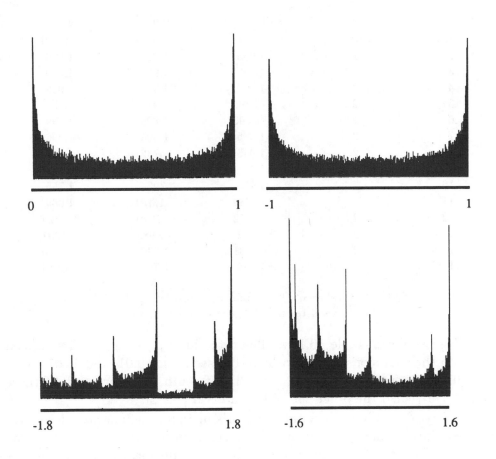

Fig. 10.6 Histograms for (a) $F_4(x) = 4x(1-x)$, (b) $G(x) = 4x^3 - 3x$,
(c) $Q_{-1.8}(x) = x^2 - 1.8$, and (d) $Q_{-1.6}(x) = x^2 - 1.6$.

histograms are clearly quite different, but they nonetheless indicate that the corresponding functions have orbits that are dense in a certain interval.

Note that histograms may be misleading: the doubling function on the unit interval is chaotic, but the computer shows (recall Experiment 3.6) that all orbits tend to zero.

Sensitive dependence on initial conditions may often be noted in rather stunning fashion using the computer, for orbits of many chaotic systems tend to separate very quickly when sensitivity is present. For example, consider again the chaotic quadratic function $Q_{-2}(x) = x^2 - 2$. We have proved that this function depends sensitively on initial conditions. We know that 0 is eventually fixed with orbit $0, -2, 2, 2, 2, \ldots$. Nearby orbits should separate

n	0	0.1	0.01	0.001
1	-2	-1.99	-1.999	-1.999
2	2	1.960	1.999	1.999
3	2	1.842	1.998	1.999
4	2	1.392	1.993	1.999
5	2	-.0597	1.974	1.999
6	2	-1.996	1.898	1.999
7	2	1.985	1.604	1.996
8	2	1.943	.5732	1.984
9	2	1.775	-1.671	1.937
10	2	1.153	.7934	1.755
11	2	-.6689	-1.370	1.080
12	2	-1.552	-.1220	-.8324
13	2	.4102	-1.985	-1.307
14	2	-1.831	1.9406	-.2916
15	2	1.355	1.766	-1.914

Table 10.1 Values of $Q^n_{-2}(x_0)$ for $x_0 = 0, 0.1, 0.01, 0.001$.
Points are listed to only three decimal places.

from this orbit. In Table 10.1 we illustrate this by listing the first fifteen points on the orbits of several nearby initial seeds, $0.1, 0.01$, and 0.001. Note how quickly these orbits separate: by the fifteenth iteration, each of these orbits has reached the other end of the interval $[-2, 2]$.

Sensitive dependence is not at all surprising in functions whose derivative is always larger than 1 in magnitude, such as in the case of the doubling function. For $x^2 - 2$, however, this result is by no means clear. We have $Q'_c(0) = 0$, so there is an interval about 0 that is contracted by Q_c. This means that any orbit that comes close to 0 is, upon the next iteration, mapped very close to $Q_c(0)$. Hence, at least in the short run, these two orbits do not separate. In the long run, however, we know that these orbits do tend to separate. Incidentally, this fact provides the answer to the essay question 18 at the end of Chapter 8.

10.4 Experiment: Feigenbaum's Constant

Goal: Using the orbit diagram, we have seen in the previous lab that the quadratic function $Q_c(x) = x^2 + c$, the logistic function $F_c(x) = cx(1 - x)$, and the sine function $c\sin(x)$ all undergo a sequence of period-doubling bifurcations as the parameter tends to the chaotic regime. We have also seen that magnifications of the orbit diagram tend to look "the same." In this experiment, we will see that there really is some truth to this: we will see that these period doubling bifurcations always occur at the same rate.

Procedure: In this experiment you will work with either the quadratic or the logistic family. We first need a definition:

Definition. Suppose x_0 is a critical point for F, that is, $F'(x_0) = 0$. If x_0 is also a periodic point of F with period n, then the orbit of x_0 is called *superstable*. The reason for this terminology is that $(F^n)'(x_0) = 0$.

In this experiment we will first determine the c-values at which either $Q_c(x) = x^2 + c$ or $F_c(x) = cx(1-x)$ have superstable cycles of periods 1, 2, 4, 8, 16, 32, and 64. Using one of the programs, experimentally determine the c-values at which your function has a superstable point of the given period. Be sure to check that this point has the correct *prime* period. You should be looking in the Period-1 window of the corresponding orbit diagram for these points.

There are a number of ways to do this. For example, you could list 1000 points on the orbit of a random initial condition. By changing the parameter, you should then search for the "exact" c for which the critical point (0 for $Q_c(x) = x^2 + c$, 0.5 for $F_c(x) = cx(1-x)$) lies on a cycle. You will usually not be able to find this parameter value exactly. However, you should find the c value for which you come closest to having the critical point periodic. Usually, this means that you find a point on the orbit within 10^{-6} of the critical point. You should find c accurate to seven decimal places, which is more or less the accuracy of the computations.

Another approach might be to compute the first 2^n points on the orbit of the critical point, and then seeing how close you come to this point. Then modify the parameter repeatedly to try to come closer to the value for which the critical point is periodic with the right period. After finding the seven c-values for your function, record these numbers in tabular form:

1. $c_0 = c$-value for period 2^0
2. $c_1 = c$-value for period 2^1
3. $c_2 = c$-value for period 2^2
4. $c_3 = c$-value for period 2^3
5. $c_4 = c$-value for period 2^4
6. $c_5 = c$-value for period 2^5
7. $c_6 = c$-value for period 2^6

Mitchell Feigenbaum

In 1975, *Mitchell Feigenbaum* (1945–) was working at the Los Alamos National Laboratory when he listened to a lecture by Stephen Smale about the period-doubling behavior of quadratic functions. Curious, he began to experiment using an old HP-65 hand-held calculator. The experiment he performed was essentially the experiment in this chapter. Not only did Feigenbaum make the observation that all functions that undergo period doubling do so at the same geometric rate, but he also drew on his background as a theoretical physicist to explain why this happens. Known as renormalization group analysis, this procedure was later used by Collet, Eckmann, and Lanford to prove the "universality" of Feigenbaum's constant. Feigenbaum is now a Professor of Physics at Rockefeller University.

Now use a calculator to compute the following ratios:

$$f_0 = \frac{c_0 - c_1}{c_1 - c_2}, f_1 = \frac{c_1 - c_2}{c_2 - c_3}, \ldots, f_4 = \frac{c_4 - c_5}{c_5 - c_6}.$$

List these numbers in tabular form, too. Do you notice any convergence? You should, at least if you have carried out the above search to enough decimal places.

Notes and Questions:

1. This number is called *Feigenbaum's constant*. It turns out that this number is "universal"—it appears whenever a typical family undergoes the period-doubling route to chaos. Compare the results of both the quadratic and logistic families. Are they the same? Try also the family $c \sin x$. What is the result here? The fact that these numbers are always the same is a remarkable result due to Feigenbaum several years ago.

2. This lab takes a long time to perform. It is useful to work with friends and divide up the tasks.

Exercises

For each of the following sets, decide whether or not the set is dense in $[0, 1]$.

1. S_1 is the set of all real numbers in $[0, 1]$ except those of the form $1/2^n$ for $n = 1, 2, 3, \ldots$.

2. S_2 is the set of all rationals in $[0, 1]$ of the form $p/2^n$, where p and n are natural numbers.

3. S_3 is the Cantor middle-thirds set.

4. S_4 is the complement of the Cantor middle-thirds set.

5. S_5 is the complement of any subset of $[0, 1]$ which has countably many elements.

For each of the following sets, decide whether or not the set is dense in Σ. Give reasons.

6. $T_1 = \{(s_0 s_1 s_2 \ldots) \mid s_4 = 0\}$.

7. T_2 is the complement of T_1.

8. $T_3 = \{(s_0 s_1 s_2 \ldots) \mid \text{the sequence ends in all 0's}\}$.

9. $T_4 = \{(s_0 s_1 s_2 \ldots) \mid \text{at most one of the } s_j = 0\}$.

10. $T_5 = \{(s_0 s_1 s_2 \ldots) \mid \text{infinitely many of the } s_j = 0\}$.

11. T_6 is the complement of T_5.

12. $T_7 = \{(s_0 s_1 s_2 \ldots) \mid \text{no two consecutive } s_j = 0\}$.

13. T_8 is the complement of T_7.

14. Find a (nontrivial) sequence of periodic points in Σ that converges to the point $(01\overline{01})$.

15. Is the orbit of the point $(01\,001\,0001\,00001\ldots)$ under σ dense in Σ?

16. Is it possible to give an example of an orbit under σ that accumulates on (that is, comes arbitrarily close to but never equals) the two fixed points of σ, but which is not dense?

17. Prove that, if $\mathbf{s} \in \Sigma$, there are sequences $\mathbf{t}$ arbitrarily close to $\mathbf{s}$ for which $d[\sigma^n(\mathbf{s}), \sigma^n(\mathbf{t})] = 2$ for all sufficiently large n.

18. Prove that the set of endpoints of removed intervals in the Cantor middle-thirds set is a dense subset of the Cantor set.

19. Let $V(x) = 2|x| - 2$. Find the fixed points of V and V^2. Compute an expression for V^3.

20. Prove that the doubling function given by

$$D(x) = \begin{cases} 2x & \text{if } x < \frac{1}{2} \\ 2x - 1 & \text{if } x \geq \frac{1}{2} \end{cases}$$

is chaotic on $[0, 1)$. Compare this result with your observations in Experiment 3.6.

21. Prove that the function

$$T(x) = \begin{cases} 2x & \text{if } x \leq \frac{1}{2} \\ 2 - 2x & \text{if } x > \frac{1}{2} \end{cases}$$

is chaotic on $[0, 1]$.

22. Use the results of the previous exercise to construct a conjugacy between T on the interval $[0, 1]$ and $G(x) = 2x^2 - 1$ on the interval $[-1, 1]$.

23. Construct a conjugacy that is valid on all of $\mathbf{R}$ between G in the previous exercise and Q_{-2}. (*Hint:* Use a linear function of the form $ax + b$.)

24. Prove that $F_4(x) = 4x(1 - x)$ is chaotic on $[0, 1]$.

25. Prove that the "tripling map" on S^1 given by $F(\theta) = 3\theta$ is chaotic.

26. Use the results of the previous exercise to prove that $G(x) = 4x^3 - 3x$ is chaotic on $[-1, 1]$.

27. Prove that

$$L(x) = \begin{cases} 3x & \text{if } x \leq \frac{1}{3} \\ -\frac{3}{2}x + \frac{3}{2} & \text{if } x > \frac{1}{3} \end{cases}$$

is chaotic on $[0, 1]$.

CHAPTER 11

Sarkovskii's Theorem

In Chapter 8 we saw that, as the quadratic function makes the transition to chaos, there seem to be many c-values where the map is chaotic, and many other c-values for which the dynamics are quite tame. In particular, in any of the windows, there seems to be at most one attracting cycle and no other dynamics. In this chapter, we will show that this is by no means the case. In particular, we will show that there is much more going on than meets the eye—particularly in the period-3 window.

11.1   Period 3 Implies Chaos

Before discussing Sarkovskii's Theorem in full generality, we will prove a very special case of this result.

The Period 3 Theorem. *Suppose $F: \mathbf{R} \to \mathbf{R}$ is continuous. Suppose also that F has a periodic point of prime period 3. Then F also has periodic points of all other periods.*

This theorem is remarkable for the simplicity of its statement. The only assumption is that F is continuous. If we find a cycle of period 3 for F, we are guaranteed that there are infinitely many other cycles for this map, with every possible period. In particular, this shows that there is much more going on in the orbit diagram for the quadratic map considered in the previous chapter. In Figure 8.5, we see only a cycle of period 3 for an interval of c-values. Somewhere in this picture there must also be a large set

James Yorke

In 1975, *James Yorke* (1941–) teamed with T.-Y. Li to publish a paper entitled "Period Three Implies Chaos."* This paper contains a proof of the Period 3 Theorem, a very special case of Sarkovskii's Theorem, which was at that time unknown in the West. The paper has become a landmark in that its title represents the first use of the word "Chaos" in the scientific literature. Yorke has made fundamental contributions in dynamical systems, particularly in the study of fractal basin boundaries and mathematical models for the spread of infectious diseases. He is currently Professor of Mathematics and Director of the Institute for Physical Sciences and Technology at the University of Maryland.

of other periodic points. We don't see them because, presumably, they are all repelling.

To prove this result we need to make two preliminary observations.

Observation 1: *Suppose $I = [a, b]$ and $J = [c, d]$ are closed intervals and $I \subset J$. If $F(I) \supset J$, then F has a fixed point in I.*

This, of course, is an immediate consequence of the Intermediate Value Theorem. Since $J \supset I$, it follows that the graph of F must cross the diagonal over I. Figure 11.1 shows that the fixed point need not be unique—there may in fact be a number of fixed points for F in I. However, the Intermediate Value Theorem guarantees that there is at least one fixed point.

Observation 2: *Suppose I and J are two closed intervals and $F(I) \supset J$. Then there is a closed subinterval $I' \subset I$ such that $F(I') = J$.*

This observation does not imply that F is one-to-one on the subinterval I'; we only claim that F maps I' onto J (see Figure 11.2). Note also that we do not assume here that $J \supset I$ as in observation 1. Now let's turn to a proof of the Period 3 Theorem.

---

* Li, T.-Y., and Yorke, J., "Period Three Implies Chaos," *American Mathematical Monthly* **82** (1975), 985-992.

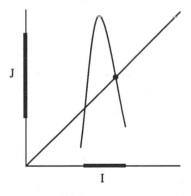

Fig. 11.1 Since $F(I) \supset J$, there is at least
one fixed point x_0 in I.

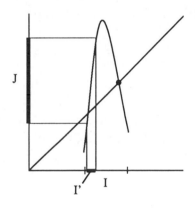

Fig. 11.2 F maps $I' \subset I$ onto J.

Proof of the Theorem: Suppose that F has a 3-cycle given by

$$a \mapsto b \mapsto c \mapsto a \mapsto \ldots .$$

If we assume that a is the leftmost point on the orbit, then there are two possibilities for the relative positions of the points on this orbit, as shown in Figure 11.3. We will assume the first case $a < b < c$; the second case is handled similarly.

Let $I_0 = [a, b]$ and $I_1 = [b, c]$. Since $F(a) = b$ and $F(b) = c$, we have $F(I_0) \supset I_1$. Similarly, since $F(c) = a$, we have $F(I_1) \supset I_0 \cup I_1$. Notice that we are making use of the assumption that F is continuous at this point.

We will first produce a cycle of period $n > 3$. Later we will handle the two special cases $n = 1$ and 2.

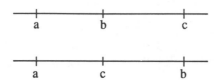

Fig. 11.3 Two possibilities for the 3-cycle $a \mapsto b \mapsto c$.

To find a periodic point of period n, we will invoke observation 2 precisely n times. First we note that, since $F(I_1) \supset I_1$, there is a closed subinterval $A_1 \subset I_1$ that satisfies $F(A_1) = I_1$. Since $A_1 \subset I_1$ and $F(A_1) = I_1 \supset A_1$, we may invoke observation 2 again to find a closed subinterval $A_2 \subset A_1$ such that $F(A_2) = A_1$. Note that, by construction, $A_2 \subset A_1 \subset I_1$ and $F^2(A_2) = I_1$.

Now continue in this fashion for $n - 2$ steps. We produce a collection of closed subintervals

$$A_{n-2} \subset A_{n-3} \subset \cdots \subset A_2 \subset A_1 \subset I_1$$

such that $F(A_i) = A_{i-1}$ for $i = 2, \ldots, n-2$ and $F(A_1) = I_1$. In particular, $F^{n-2}(A_{n-2}) = I_1$ and $A_{n-2} \subset I_1$.

Now, since $F(I_0) \supset I_1 \supset A_{n-2}$, there is also a closed subinterval $A_{n-1} \subset I_0$ such that $F(A_{n-1}) = A_{n-2}$. Finally, since $F(I_1) \supset I_0 \supset A_{n-1}$, there is another closed subinterval $A_n \subset I_1$ such that $F(A_n) = A_{n-1}$. Putting this all together, we find

$$A_n \xrightarrow{F} A_{n-1} \xrightarrow{F} \cdots \xrightarrow{F} A_1 \xrightarrow{F} I_1$$

with $F(A_i) = A_{i-1}$ so that $F^n(A_n) = I_1$ (see Figure 11.4). But $A_n \subset I_1$, so we may use observation 1 to conclude that there is a point $x_0 \in A_n$ that is fixed by F^n. Hence x_0 has period n. We claim that x_0 has prime period n.

To see this, note that $F(x_0) \in A_{n-1} \subset I_0$, but $F^i(x_0) \in I_1$ for $i = 2, \ldots, n$. So the first iterate of x_0 lies in I_0 but all others lie in I_1. This proves that x_0 has period $\geq n$, so x_0 has prime period n.*

The final cases are $n = 1$ and 2. These are handled by noting that $F(I_1) \supset I_1$, so there is a fixed point in I_1. Similarly, $F(I_0) \supset I_1$ and $F(I_1) \supset I_0$. So there is a 2-cycle that hops back and forth between I_0 and I_1, using the above argument. This completes the proof.

---

*   Note that x_0 cannot lie in $I_0 \cap I_1 = \{b\}$, for in this case we would have $b \in A_{n-1}$ so $F(b) = c \notin I_0$.

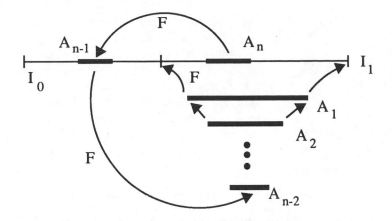

Fig. 11.4 Construction of the A_i.

11.2   Sarkovskii's Theorem

Sarkovskii's Theorem, first proved in 1964, is an incredibly powerful and beautiful strengthening of the Period 3 Theorem. To state the theorem, we first list all of the natural numbers in the following strange order:

$$3, 5, 7, 9, \ldots$$
$$2 \cdot 3, 2 \cdot 5, 2 \cdot 7, \ldots$$
$$2^2 \cdot 3, 2^2 \cdot 5, 2^2 \cdot 7, \ldots$$
$$2^3 \cdot 3, 2^3 \cdot 5, 2^3 \cdot 7, \cdots$$
$$\vdots$$
$$\ldots, 2^n, \ldots, 2^3, 2^2, 2^1, 1.$$

This is known as the Sarkovskii ordering of the natural numbers.

Sarkovskii's Theorem. *Suppose $F: \mathbf{R} \to \mathbf{R}$ is continuous. Suppose that F has a periodic point of period n and that n precedes k in the Sarkovskii ordering. Then F also has a periodic point of prime period k.*

Again note the simplicity of the hypotheses and power of the conclusion of this theorem. From Sarkovskii's result, we know that a continuous function that has a periodic point of period 6 must also have cycles of all periods

A. N. Sarkovskii

In 1964, A. N. Sarkovskii published a short paper that includes the theorem that now bears his name. Originally published in Russian*, this paper remained unknown in the West until Li and Yorke published their result ten years later. Now one of the leading figures in the study of one-dimensional dynamical systems, Sarkovskii was one of the first mathematicians to suggest that iteration of functions of a real variable is a subject in its own right. He is now Professor of Mathematics at the Institute for Mathematics in Kiev, Ukraine.

---

except possibly an odd number greater than one. Similarly, if F has period $56 = 7 \cdot 2^3$, then F must also have periods $72, 88, 104, \ldots, 8, 4, 2$, and 1—the entire tail of the Sarkovskii list.

Perhaps even more amazing is the fact that the converse of Sarkovskii's Theorem is also true:

Theorem. *There is a continuous function* $F\colon \mathbf{R} \to \mathbf{R}$ *which has a cycle of period* n, *but no cycles of any period that precedes* n *in the Sarkovskii ordering.*

Remarks:

1. Sarkovskii's Theorem is also true if $F\colon I \to I$ is continuous, where I is a closed interval of the form $[a, b]$. We simply extend F to the entire real line by defining $F(x) = F(a)$ if $x < a$, and $F(x) = F(b)$ if $x > b$. The resulting extension is clearly a continuous function.
2. Note that the Period 3 Theorem is an immediate corollary of Sarkovskii's Theorem: 3 heads the list of numbers in Sarkovskii's ordering. Hence a continuous function with a 3-cycle has all other periods.
3. Note also that the numbers of the form 2^n form the tail of the Sarkovskii ordering. Thus, if F has only finitely many periodic points, then they all must have periods of a power of 2. This explains in part why we always see period doubling at the outset when a family of functions makes the transition from simple dynamics to chaos.

---

* "Coexistence of Cycles of a Continuous Map of a Line into Itself," *Ukrain. Mat. Z.* **16** (1964), 61-71 (In Russian).

4. Sarkovskii's Theorem is not true if the "space" in question is anything but the real line or interval. For example, the function on the circle that just rotates all points by a fixed angle $2\pi/n$ has periodic points of period n and no other periods.

We will not present the full proof of Sarkovskii's Theorem. The proof is not difficult: it basically involves repeated applications of the two observations we made in the previous section. However, the two intervals I_0 and I_1 used in the proof of the Period 3 Theorem must be replaced by $n - 1$ intervals. This means that the "bookkeeping" becomes considerably more involved. See [Devaney, p. 63] for more details. Instead, we will give a flavor of the full proof by dealing with several special cases.

Case 1: Period $k \Rightarrow$ Period 1. This follows from the Intermediate Value Theorem. Suppose $x_1, \ldots, x_k$ lie on the k-cycle, with

$$x_1 < x_2 < \cdots < x_k.$$

Now $F(x_1)$ must be one of the x_i with $i > 1$, and $F(x_k)$ is similarly an x_i with $i < k$. Thus $F(x_1) = x_i$ for some $i > 1$, we have $F(x_1) - x_1 > 0$. Similarly $F(x_k) - x_k < 0$. Therefore, there is an x between x_1 and x_k with $F(x) - x = 0$, which gives us a fixed point.

Case 2: Period 4 $\Rightarrow$ Period 2. This case is more complicated. Suppose x_1, x_2, x_3, x_4 form the 4-cycle with

$$x_1 < x_2 < x_3 < x_4.$$

Choose a point a between x_2 and x_3. Then there are two cases. The first case occurs if both $F(x_1) > a$ and $F(x_2) > a$. Then we must have $F(x_3) < a$ and $F(x_4) < a$. Let $I_0 = [x_1, x_2]$ and $I_1 = [x_3, x_4]$. Since the x_i's are permuted, it follows that $F(I_1) \supset I_0$ and $F(I_0) \supset I_1$. Our two observations in Section 11.1 then guarantee that there is a 2-cycle that hops between I_0 and I_1.

The other case occurs when one of x_1 or x_2 is mapped to the right of a but the other is not. For definiteness, suppose $F(x_1) > a$ and $F(x_2) < a$. Consequently, we must have $F(x_2) = x_1$. Let $I_0 = [x_2, x_3]$ and $I_1 = [x_1, x_2]$ (see Figure 11.5). Then we have $F(I_0) \supset I_1$ and $F(I_1) \supset I_0 \cup I_1$. This is the same situation we encountered in the proof of the Period 3 Theorem. Hence there is a cycle of period 2 (and, in fact, a cycle of *any* period). The other

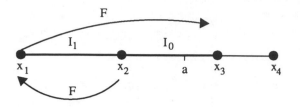

Fig. 11.5 Period 4 $\Rightarrow$ Period 2.

possibility is handled in similar fashion, except that we must make different choices for I_0 and I_1.

Case 3: Period $2^n \Rightarrow$ Period 2^k when $n > k$. The previous two cases take care of the case when $n = 1$ or 2, so we assume $n \geq 3$. Let $\ell = 2^{n-2}$ and consider $G(x) = F^\ell(x)$. The cycle of period 2^n for F is a cycle of period 4 for G. It follows from the previous case that G has a 2-cycle. But this 2-cycle for G is a 2^{n-1}-cycle for F.

Now let us give several examples of the converse of Sarkovskii's Theorem.

Example. Figure 11.6 shows a sketch of the graph of a piecewise linear* function defined on the interval $1 \leq x \leq 5$. Note that

$$F(1) = 3$$
$$F(3) = 4$$
$$F(4) = 2$$
$$F(2) = 5$$
$$F(5) = 1$$

so that we have a 5-cycle:

$$1 \mapsto 3 \mapsto 4 \mapsto 2 \mapsto 5 \mapsto 1\ldots.$$

To see that F has no periodic point of period 3, we assume that there is such a point. From the graph, we see that

$$F([1,2]) = [3,5]$$
$$F([3,5]) = [1,4]$$
$$F([1,4]) = [2,5].$$

---

*   Piecewise linear simply means that the graph of the function is a finite collection of straight lines.

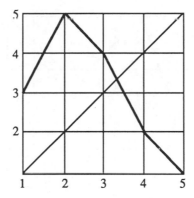

Fig. 11.6 *F* has period 5 but not period 3.

Hence, $F^3([1,2]) = [2,5]$ so $F^3([1,2]) \cap [1,2] = \{2\}$, which has period 5. Hence F^3 has no fixed points in $[1,2]$.

Similar arguments show that there are no 3-cycles in either of the intervals $[2,3]$ or $[4,5]$. We cannot use the same argument in the interval $[3,4]$, since F itself has a fixed point within this interval. However, we note that

$$F: [3,4] \to [2,4]$$

is a decreasing function. Also,

$$F: [2,4] \to [2,5]$$

and

$$F: [2,5] \to [1,5]$$

are also decreasing. The composition of an odd number of decreasing functions is decreasing. Hence

$$F^3: [3,4] \to [1,5]$$

is decreasing. Thus the graph of F^3 on $[3,4]$ meets the diagonal over $[3,4]$ in exactly one point. This point must be the fixed point of F. Therefore F has no 3-cycles in $[3,4]$ either. Consequently, this function has a period 5 point but no period 3 point.

Example. Figure 11.7 shows the graph of a piecewise linear function that has a 7-cycle given by

$$1 \mapsto 4 \mapsto 5 \mapsto 3 \mapsto 6 \mapsto 2 \mapsto 7 \mapsto 1 \mapsto \dots.$$

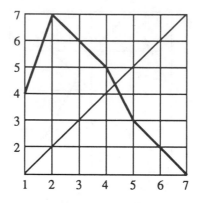

Fig. 11.7 F has a 7-cycle but no 5-cycle.

Arguments similar to those in the previous example show that F has no 5-cycle.

11.3   The Period-3 Window

Now let's return to our friend, the quadratic family $Q_c(x) = x^2 + c$. Recall from Chapter 8 that there is an interval of c-values for which Q_c apparently has an attracting 3-cycle. In the corresponding window in the orbit diagram we saw only this 3-cycle—not all of the other periodic points whose existence Sarkovskii's Theorem guarantees. In this and the next section we will locate these points and show how to use symbolic dynamics to analyze them.

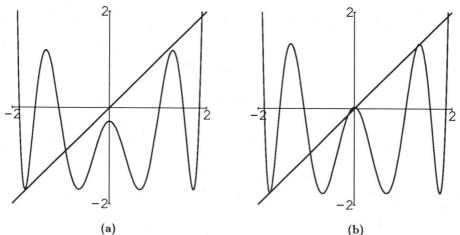

(a)   (b)

Fig. 11.8 Graphs of Q_c^3 for (a) $c = -1.7$, (b) $c = -1.76$.

Using the orbit diagram for Q_c, one may check that the period 3 window first opens for $c \approx -1.75$. Figure 11.8 shows the graphs of Q_c^3 for two c-values near -1.75. These graphs show that this family undergoes a saddle-node bifurcation as c decreases through $-1.75\ldots$.

For the remainder of this section we will consider the specific parameter value $c = -1.7548777\ldots$. This value is chosen so that 0 lies on the attracting 3-cycle

$$0 \mapsto c \mapsto c^2 + c \mapsto 0.$$

So c is the nonzero root of the equation

$$0 = Q_c^3(0) = (c^2 + c)^2 + c$$

(see Figure 11.9). For simplicity of notation, we will write $Q = Q_{-1.7548777\ldots}$ for the remainder of this and the next section. This c-value is called a *superstable* parameter since the derivative of Q^3 along the attracting 3-cycle is 0.

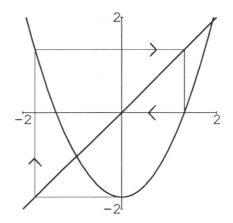

Fig. 11.9 The attracting 3-cycle for Q contains 0.

Figure 11.10 shows the graph of Q^3. Although difficult to see from afar, this graph has eight fixed points, the two fixed points of Q, the attracting 3-cycle, and another repelling 3-cycle which we denote by

$$\alpha \mapsto \beta \mapsto \gamma \mapsto \alpha$$

with $\gamma < \beta < \alpha$. From this graph we see that each of α, β, and γ has a nearby point $\hat{\alpha}, \hat{\beta}$, and $\hat{\gamma}$ respectively that satisfies

$$Q^3(\hat{\alpha}) = \alpha$$
$$Q^3(\hat{\beta}) = \beta$$
$$Q^3(\hat{\gamma}) = \gamma.$$

Figure 11.11, the enlarged graph of Q^3 in the vicinity of α, clearly shows the relative positions of α and $\hat{\alpha}$.

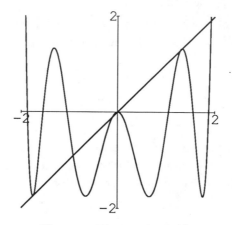

Fig. 11.10 The graph of Q^3.

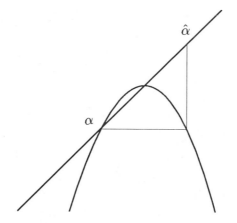

Fig. 11.11 Magnification of the graph of Q^3 showing $\alpha, \hat{\alpha}$.

By graphical analysis we see that any point in the interval $(\alpha, \hat{\alpha})$ has an orbit that tends under Q^3 to the attracting fixed point for Q^3 in this interval. The same is true in the intervals $(\beta, \hat{\beta})$ and $(\hat{\gamma}, \gamma)$. Hence we know the fate of any orbit that enters one of these intervals—it is attracted to the attracting 3-cycle. In particular, there are no other cycles in these intervals.

In Figure 11.12 are superimposed the graphs of Q and Q^3. These graphs show that

$$Q(\hat{\alpha}) = \hat{\beta}$$
$$Q(\hat{\gamma}) = \hat{\alpha},$$

but

$$Q(\hat{\beta}) = \gamma.$$

We also have

$$Q([\alpha, \hat{\alpha}]) = [\beta, \hat{\beta}]$$
$$Q([\hat{\gamma}, \gamma]) = [\alpha, \hat{\alpha}],$$

but

$$Q([\beta, \hat{\beta}]) \subset [\hat{\gamma}, \gamma]$$

since $Q(\beta) = Q(\hat{\beta}) = \gamma$.

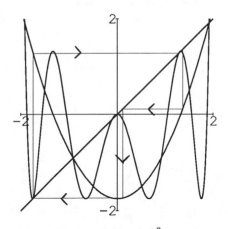

Fig. 11.12 Graphs of Q and Q^3 superimposed.

Let $I_0 = [\hat{\beta}, \alpha]$ and $I_1 = [\gamma, \beta]$. From the graph we see that Q maps I_0 in one-to-one fashion onto I_1 while Q takes I_1 into $I_0 \cup (\beta, \hat{\beta}) \cup I_1$. In particular, just as in the proof of the Period 3 Theorem, we have

$$Q(I_0) \supset I_1$$
$$Q(I_1) \supset I_0 \cup I_1.$$

Thus the proof of that theorem shows that we have periodic points of all periods in $I_0 \cup I_1$.

In fact, all cycles (except p_+ and the attracting 3-cycle) lie in $I_0 \cup I_1$. The only other possible locations for cycles would be in the intervals $(\hat{a}, p_+)$ and $(-p_+, \hat{\gamma})$. But graphical analysis shows that any point in these intervals has an orbit that eventually enters the interval $[\hat{\gamma}, \hat{a}]$. Once in this interval, the orbit may never escape, since $Q([\hat{\gamma}, \hat{a}]) \subset [\hat{\gamma}, \hat{a}]$. Indeed, $Q([\hat{\gamma}, \hat{a}]) \subset [c, \hat{a}]$ and $\hat{\gamma} < c$. Figure 11.13 sketches the image of $[-p_+, p_+]$, showing the images of each of the important subintervals.

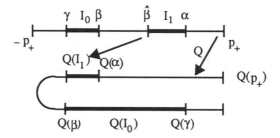

Fig. 11.13 Image of $[-p_+, p_+]$ under Q.

Thus, to understand the dynamics of Q completely, we need only to analyze the behavior of Q in the intervals I_0 and I_1. A slight variation of symbolic dynamics allows us to accomplish this.

11.4   Subshifts of Finite Type

We continue using the notation $Q(x) = x^2 - 1.7548777\ldots$. To understand the dynamics of Q completely, we need to analyze the orbits that remain for all time in $I_0 \cup I_1$. Recall that if an orbit of a point in $I_0 \cup I_1$ ever leaves these intervals, then that orbit must enter $(\beta, \hat{\beta})$ and thus be attracted to the attracting 3-cycle. Therefore, we need to understand the dynamics of Q on the set Λ defined by

$$\Lambda = \{x \in I_0 \cup I_1 \mid Q^n(x) \in I_0 \cup I_1 \text{ for all } n\}.$$

For a point x in Λ we define the itinerary of x, $S(x)$, much as we did in Chapter 9:

$$S(x) = (s_0 s_1 s_2 \ldots),$$

where $s_j = 0$ or 1 for all j, and $s_j = 0$ if $Q^j(x) \in I_0$, $s_j = 1$ if $Q^j(x) \in I_1$. There is one significant difference between the possible itineraries for Q and those for the quadratic map Q_c with $c < -2$ discussed earlier. Suppose x has itinerary $(s_0 s_1 s_2 \ldots)$ and $Q^j(x) \in I_0$ so that $s_j = 0$. Since $Q(I_0) = I_1$, it follows that $Q^{j+1}(x) \in I_1$. Hence $s_{j+1} = 1$. Therefore we see that, if any entry of the itinerary of x is 0, the entry that immediately follows must be 1. That is, not all sequences of 0's and 1's are allowable itineraries for Q in Λ. For example, neither $(000\ldots)$ nor $(001001001\ldots)$ correspond to itineraries of points in Λ. Equivalently, there is no fixed point that resides in I_0, so $(000\ldots)$ is not allowable. From Figure 11.12 we see that Q has a fixed point in $(\gamma, \beta) \subset I_1$. That is why $(111\ldots)$ is an allowable itinerary.

Mimicking what we did in Chapter 9, we denote by Σ' the set of all itineraries that satisfy the above condition; that is,

$$\Sigma' = \{(s_0 s_1 s_2 \ldots) \mid s_j = 0 \text{ or } 1, \text{ but if } s_j = 0 \text{ then } s_{j+1} = 1\}.$$

Note that $\Sigma' \subset \Sigma$, the set of all possible sequences of 0's and 1's. So we may use the same distance function introduced in Chapter 9 to measure distances in Σ'. In particular, we may use this to show that Σ' is a closed subset of Σ. We are familiar with the notion of open and closed intervals in $\mathbf{R}$. More generally, we may define open and closed sets as follows.

Definition. Suppose Y is a subset of a set X equipped with a metric d. We say that Y is an *open subset* of X if for any $y \in Y$ there is an $\epsilon > 0$ such that, if $d[x, y] < \epsilon$, then $x \in Y$.

That is, open sets have the property that we can always find a small open "ball" in Y—all points within distance ϵ—around any point in Y.

Definition. A subset $V \subset X$ is a *closed set* if the complement of V is an open subset of X.

Thus, to show that $\Sigma' \subset \Sigma$ is a closed subset, we must verify that its complement is open. To see this, choose a point $\mathbf{s} = (s_0 s_1 s_2 \ldots)$ in the complement of Σ'. We must produce an $\epsilon > 0$ such that all points within ϵ of $\mathbf{s}$ also lie in the complement of Σ'.

Since $\mathbf{s} \notin \Sigma'$, there must be at least one pair of adjacent 0's in the sequence $(s_0 s_1 s_2 \ldots)$. Suppose k is such that $s_k = s_{k-1} = 0$. Let's choose $\epsilon <$

$1/2^k$. If $\mathbf{t} \in \Sigma$ satisfies $d[\mathbf{s}, \mathbf{t}] < \epsilon$, then the Proximity Theorem guarantees that $t_0 = s_0, \ldots, t_k = s_k$. In particular, $t_k = t_{k-1} = 0$. Therefore $\mathbf{t}$ lies in the complement of Σ' and we have found a small "ball" about $\mathbf{s}$ in the complement of Σ'. This proves that Σ' is closed. So we have

Proposition. $\Sigma' \subset \Sigma$ *is a closed subset.*

The shift map σ also makes sense on Σ'. For if $\mathbf{s}$ is a sequence with no pair of adjacent 0's, then $\sigma(\mathbf{s})$ also has this property. The natural question is what is the relationship between Q on Λ and the shift σ restricted to Σ'. As you might expect, the itinerary function $S: \Lambda \to \Sigma'$ provides the answer.

Theorem. *The itinerary function* $S: \Lambda \to \Sigma'$ *is a conjugacy between* $Q: \Lambda \to \Lambda$ *and* $\sigma: \Sigma' \to \Sigma'$.

We will not provide the details of the proof since it is essentially the same as that of Conjugacy Theorem in Chapter 9. The one important difference is that $|Q'(x)|$ is not everywhere larger than 1 on $I_0 \cup I_1$. However, one may check that $|(Q^k)'(x)| > 1$ for some k and all $x \in I_0 \cup I_1$, and this is sufficient to complete the proof. Full details may be found in [Devaney, pp. 98-99].

The shift map is also chaotic on Σ'. Notice that this needs proof; just because periodic points are dense in Σ, it does not necessarily follow that they are dense in a subset of Σ. Similarly, how do we know that there is a dense orbit in Σ'? We will leave the proofs of these facts as exercises since these proofs are similar in spirit to those in Chapter 8.

Theorem. *The shift map* $\sigma: \Sigma' \to \Sigma'$ *is chaotic.*

From Sarkovskii's theorem we know that Q has periodic points of all periods in Λ, but this result gives us no indication of how many cycles Q has. For this information, we need to make use of the symbolic dynamics.

Let Per_n denote the set of sequences in Σ' that are fixed by σ^n. Note that Per_1 contains only one sequence, $(11\overline{1})$, while Per_2 contains $(01\overline{01})$, $(10\overline{10})$, and $(11\overline{1})$. Our goal is to find a formula for the number of sequences in Per_n for all n.

There are three distinct types of sequences in Per_n, namely

$$A_n = \{(\overline{s_0 \cdots s_{n-1}}) \in \text{Per}_n \,|\, s_0 = 1, s_{n-1} = 0\}$$
$$B_n = \{(\overline{s_0 \cdots s_{n-1}}) \in \text{Per}_n \,|\, s_0 = 0, s_{n-1} = 1\}$$
$$C_n = \{(\overline{s_0 \cdots s_{n-1}}) \in \text{Per}_n \,|\, s_0 = 1 = s_{n-1}\}.$$

Note that a repeating sequence of the form $(\overline{0s_1 \ldots s_{n-2}0})$ does not lie in Σ'.

Let $\#\mathrm{Per}_n$ denote the number of points in Per_n. Then we have

$$\#\mathrm{Per}_n = \#A_n + \#B_n + \#C_n,$$

since A_n, B_n, and C_n are mutually exclusive.

To determine $\#\mathrm{Per}_n$ we will show that there is a one-to-one correspondence between Per_{n+2} and $\mathrm{Per}_{n+1} \cup \mathrm{Per}_n$. This will prove

Theorem. $\#\mathrm{Per}_{n+2} = \#\mathrm{Per}_{n+1} + \#\mathrm{Per}_n$ *for* $n > 0$.

Proof. Choose any $\mathbf{s} = (\overline{s_0 s_1 \ldots s_{n+1}}) \in \mathrm{Per}_{n+2}$. We will associate a unique sequence in either Per_{n+1} or Per_n to $\mathbf{s}$. If $s_0 = s_{n+1}$ then we must have $s_0 = s_{n+1} = 1$, since adjacent 0's are not allowed. Now s_n may be either 0 or 1, so $\mathbf{s} \in \mathrm{Per}_{n+2}$ with $s_0 = s_{n+1} = 1$ determines a repeating sequence of length $n + 1$, namely $(\overline{1s_1 \ldots s_n})$, which lies in either A_{n+1} or C_{n+1}.

On the other hand, if $\mathbf{s} \in \mathrm{Per}_{n+2}$ but $s_0 \neq s_{n+1}$, then we have two cases. First, if $s_0 = 1$ and $s_{n+1} = 0$, then $s_n = 1$. So s_{n-1} may be either 0 or 1. Thus $\mathbf{s}$ determines a unique sequence $(\overline{1s_1 \ldots s_{n-1}})$ in either A_n or C_n.

Finally, if $s_0 = 0$ and $s_{n+1} = 1$, then s_n may be either 0 or 1. If $s_n = 0$ then $\mathbf{s}$ determines $(\overline{0s_1 \ldots s_{n-1}})$, which lies in B_n since $s_{n-1} = 1$. If $s_n = 1$ then $\mathbf{s}$ determines $(\overline{0s_1 \ldots s_{n-1}1})$ in B_{n+1}.

Now recall that $\mathrm{Per}_n = A_n \cup B_n \cup C_n$. Thus we may associate a unique sequence in either Per_{n+1} or Per_n to any sequence in Per_{n+2}. Reversing the above procedure yields the converse. This completes the proof.

This theorem allows us to determine recursively the number of periodic points in Λ. We have already seen that $\#\mathrm{Per}_1 = 1$ and $\#\mathrm{Per}_2 = 3$. Then the recursive relation yields

$$\#\mathrm{Per}_3 = 4$$
$$\#\mathrm{Per}_4 = 7$$
$$\#\mathrm{Per}_5 = 11$$
$$\#\mathrm{Per}_6 = 18$$

and so forth.

We remark that the recursive relation

$$\#\mathrm{Per}_{n+2} = \#\mathrm{Per}_{n+1} + \#\mathrm{Per}_n$$

generates the well-known Fibonacci sequence when $\#\text{Per}_1 = 1$ and $\#\text{Per}_2 = 1$. So the classical Fibonacci sequence is

$$1, 1, 2, 3, 5, 8, 13 \ldots,$$

and ours is a slight variation on this theme.

The shift map σ on Σ' is called a *subshift of finite type*. It is a subshift since $\Sigma' \subset \Sigma$. It is of finite type since it is determined by only finitely many conditions on the entries of sequences in Σ'. In our case, the only condition is that 0 may not follow 0.

More generally, subshifts of finite type occur as subshifts of the shift on N symbols. Let Σ_N denote the set of all sequences whose entries are $0, 1, \ldots, N - 1$. See the exercises following Chapter 9 regarding this set. A subshift of finite type is defined by prescribing which digits are allowed to follow given digits in an allowable sequence. This may be visualized most effectively as a directed graph with vertices $0, 1, \ldots, N - 1$. A directed graph consists of arrow passing from vertices to vertices with at most one arrow going from vertex i to vertex j. These arrows are directed, so there may be another arrow going from j to i. The allowable sequences are then all infinite paths through this graph. For example, the subshift of finite type corresponding to the period-3 window in the orbit diagram corresponds to the graph in Figure 11.14a. Figure 11.14b depicts the directed graph corresponding to the full shift on 2 symbols.

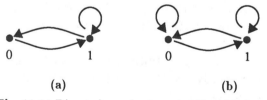

(a)   (b)

Fig. 11.14 Directed graphs for subshifts of finite type.

Subshifts of finite type need not have chaotic dynamics. For example, the directed graph in Figure 11.15 corresponds to the subshift for which 0 may follow 0 or 1, but 1 may only follow 1.

Note that the only allowable sequences in this subshift are those of the form

$$(000\ldots)$$
$$(111\ldots)$$
$$(111\ldots1000\ldots).$$

Fig. 11.15 A nonchaotic subshift of finite type.

Thus there are only two fixed points for this dynamical system, no other periodic points, and no dense orbit.

Exercises

1. Can a continuous function on **R** have a periodic point of period 48 and not one of period 56? Why?

2. Can a continuous function on **R** have a periodic point of period 176 but not one of period 96? Why?

3. Give an example of a function $F: [0,1] \to [0,1]$ that has a periodic point of period 3 and *no* other periods. Can this happen?

4. The graphs in Figure 11.16 each have a cycle of period 4 given by $\{0,1,2,3\}$. One of these functions has cycles of all other periods, and one has only periods 1, 2, and 4. Identify which function has each of these properties.

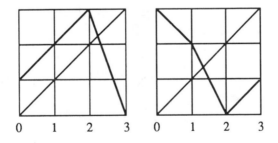

Fig. 11.16 Two graphs with period 4.

5. Suppose a continuous function F has a cycle of period $n \geq 3$ given by $a_1 < a_2 < \cdots < a_n$. Suppose that F permutes the a_i according to the rule $a_1 \mapsto a_2 \mapsto \cdots a_n \mapsto a_1$. What can you say about other cycles for F?

6. Consider the piecewise linear graph in Figure 11.7. Prove that this function has a cycle of period 7 but not period 5.

7. Consider the graph in Figure 11.17a. Prove that this function has a cycle of period 6 but no cycles of any odd period.

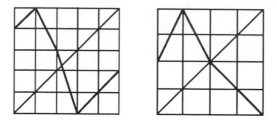

Fig. 11.17a   Fig. 11.17b

8. Consider the function whose graph is displayed in Fig. 11.17b. Prove that this function has cycles of all even periods but no odd periods (except 1).

9. Consider the subshift of finite type $\Sigma' \subset \Sigma$ determined by the rules 1 may follow 0 and both 0 and 1 may follow 1, as discussed in Section 11.4.
    a. Prove that periodic points for σ are dense in Σ'.
    b. Prove that there is a dense orbit for σ in Σ'

The following four problems deal with the subshift of Σ_3, the space of sequences of 0's, 1's, and 2's, determined by the rules that 1 may follow 0, 2 may follow 1, and 0, 1, or 2 may follow 2.

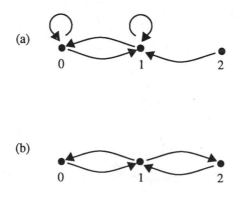

Fig. 11.18

10. Is this subset of Σ_3 closed?

11. Are periodic points dense for this subshift?

12. Is there a dense orbit for this subshift?

13. How many periodic points of periods 2, 3, and 4 satisfy these rules?

14. Construct a subshift of finite type in Σ_3 which has a period 3 cycle but no fixed or period 2 points.

15. Discuss the dynamics of the subshift given by the directed graph in Figure 11.18a. Are periodic points dense for this subshift? Is there a dense orbit? How many periodic points of period n does this subshift have?

16. Discuss the dynamics of the subshift given by the directed graph in Figure 11.18b. Are periodic points dense for this subshift? Is there a dense orbit?

CHAPTER 12

The Role of the Critical Orbit

In previous sections we have seen that simple dynamical systems such as the quadratic family may possess infinitely many periodic points. But, as we saw in the orbit diagram, very few of them appear to be attracting. In this chapter we will find out why this is so.

12.1 The Schwarzian Derivative

The Schwarzian derivative is one of the stranger tools in dynamics. Although this derivative has a venerable history in the field of complex analysis, but it was only introduced into the study of dynamical systems in 1978. Functions with negative Schwarzian derivatives have very interesting dynamical properties that simplify their analysis.

Definition. The *Schwarzian derivative* of a function F is

$$SF(x) = \frac{F'''(x)}{F'(x)} - \frac{3}{2}\left(\frac{F''(x)}{F'(x)}\right)^2.$$

Many functions have negative Schwarzian derivatives. For example, the quadratic family $Q_c(x) = x^2 + c$ satisfies $SQ_c(x) = -3/(2x^2)$. Note that $SQ_c(x) < 0$ for all x, including the critical point $x = 0$ where we may define $SQ_c(0) = -\infty$, since $\lim_{x \to 0} SQ_c(x) = -\infty$. Also,

$$S(e^x) = -\frac{1}{2} < 0$$

$$S(\sin x) = -1 - \frac{3}{2}\tan^2 x < 0$$

$$S(kx(1-x)) = \frac{-6}{(1-2x)^2} < 0.$$

Note that, in each case, the Schwarzian derivative at critical points is $-\infty$. We write $SF < 0$ whenever $SF(x) < 0$ for all x.

Many polynomials have negative Schwarzian derivatives, as the following proposition shows.

Proposition. *Suppose $P(x)$ is a polynomial and all roots of $P'(x)$ are real and distinct. Then $SP < 0$.*

Proof: Since $P'(x)$ has real and distinct roots, we may write

$$P'(x) = \alpha(x - a_1) \cdot \ldots \cdot (x - a_N).$$

Hence

$$\log P'(x) = \log \alpha + \sum_{i=1}^{N} \log(x - a_i).$$

Differentiating via the Chain Rule, we find

$$\frac{P''(x)}{P'(x)} = \sum_{i=1}^{N} \frac{1}{x - a_i}.$$

Differentiating again via the Quotient Rule, we find

$$\frac{P'''(x)P'(x) - (P''(x))^2}{(P'(x))^2} = \frac{P'''(x)}{P'(x)} - \left(\frac{P''(x)}{P'(x)}\right)^2$$

$$= -\sum_{i=1}^{N} \frac{1}{(x - a_i)^2}.$$

Hence

$$SP(x) = \frac{P'''(x)}{P'(x)} - \left(\frac{P''(x)}{P'(x)}\right)^2 - \frac{1}{2}\left(\frac{P''(x)}{P'(x)}\right)^2$$

$$= -\sum_{i=1}^{N} \frac{1}{(x - a_i)^2} - \frac{1}{2}\left(\sum_{i=1}^{N} \frac{1}{x - a_i}\right)^2$$

$$< 0.$$

This completes the proof.

The main reason for the importance of negative Schwarzian derivatives is the fact that this property is preserved by composition of functions and consequently by iteration.

Chain Rule for Schwarzian Derivatives. *Suppose F and G are functions. Then*

$$S(F \circ G)(x) = SF(G(x)) \cdot (G'(x))^2 + SG(x).$$

Proof: Using the Chain Rule for ordinary derivatives, we compute

$$(F \circ G)'(x) = F'(G(x)) \cdot G'(x)$$
$$(F \circ G)''(x) = F''(G(x)) \cdot (G'(x))^2 + F'(G(x)) \cdot G''(x)$$

Differentiating once more,

$$(F \circ G)'''(x) = F'''(G(x)) \cdot (G'(x))^3 + 3F''(G(x)) \cdot G''(x) \cdot G'(x)$$

$$+ F'(G(x)) \cdot G'''(x).$$

After a hefty dose of algebra, we find the result.

Corollary. *Suppose SF < 0 and SG < 0. Then S(F∘G) < 0. In particular, if SF < 0, then SF^n < 0.*

Proof: We have $SF(G(x)) < 0$ and $SG(x) < 0$ for all x. Hence, by the Chain Rule for Schwarzian Derivatives,

$$S(F \circ G)(x) = SF(G(x)) \cdot (G'(x))^2 + SG(x) < 0.$$

It is difficult to see geometrically what the property of negative Schwarzian derivative means. However, the following result gives an indication of the kind of graphs that cannot occur for functions with negative Schwarzian derivatives.

Schwarzian Min-Max Principle. *Suppose $SF < 0$. Then F' cannot have a positive local minimum or a negative local maximum.*

Proof: Suppose x_0 is a critical point of F'. That is, $F''(x_0) = 0$. Suppose also that $F'(x_0) \neq 0$. Then we have

$$SF(x_0) = \frac{F'''(x_0)}{F'(x_0)} < 0.$$

Now let's apply the "second derivative test" from calculus, not to F, but rather to F'. (Be careful—this can be confusing!)

If F' has a positive local minimum at x_0, then its second derivative $F'''(x_0)$ must be non-negative. Since $F'(x_0) > 0$, we have

$$\frac{F'''(x_0)}{F'(x_0)} \geq 0.$$

This contradicts $SF < 0$.

Similarly, if F' has a negative local maximum at x_0, then $F'''(x_0) \leq 0$ and $F'(x_0) < 0$, again yielding a contradiction. This concludes the proof.

As a consequence of this proposition, we see that the graph shown in Figure 12.1a cannot occur for a function with negative Schwarzian derivative. Indeed, there are two points a and b where the derivative is 1. In between, the slope is less than 1, but never negative. So F' must have a positive local minimum between a and b. This is impossible by the Schwarzian Min-Max Principle. Figure 12.1b displays a graph where $F'(x)$ has a negative local maximum between a and b, each of which has derivative equal to -1. Again, this kind of graph cannot occur if $SF < 0$.

12.2   The Critical Point and Basins of Attraction

In this section we investigate how the assumption of negative Schwarzian derivative severely limits the kinds of dynamical behavior that may occur. We will show that each attracting periodic orbit of such a function must attract at least one critical point of the function. We begin with several definitions.

Definition. Suppose x_0 is an attracting fixed point for F. The *basin of attraction* of x_0 is the set of all points whose orbits tend to x_0. The *immediate*

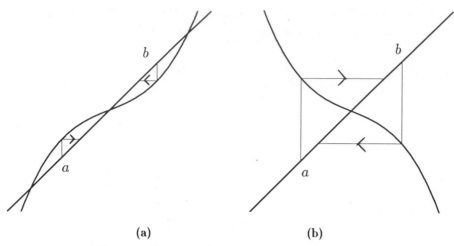

(a)   (b)

Fig. 12.1 These graphs are impossible if $SF < 0$.

basin of attraction of x_0 is the largest interval containing x_0 that lies in the basin of attraction.

Basins of attraction for attracting cycles of period n are defined using F^n instead of F. The Attracting Fixed Point Theorem (chapter 5) guarantees that attracting fixed points and cycles have immediate basins of attraction.

In general, but not always, the immediate basin of attraction of an attracting fixed point is smaller than the full basin. For $F(x) = x^2$, the immediate basin of attraction of the fixed point 0 is $(-1, 1)$. This is also the full basin of attraction. For $C(x) = \pi \cos x$, there is an attracting fixed point at $-\pi$, but its basin of attraction is much larger than the immediate basin of attraction (Figure 12.2).

Our main result in this section is the following.

Theorem. *Suppose $SF < 0$. If x_0 is an attracting periodic point for F, then either the immediate basin of attraction of x_0 extends to $+\infty$ or $-\infty$, or else there is a critical point of F whose orbit is attracted to the orbit of x_0.*

This theorem explains why we see at most one attracting periodic orbit for the quadratic family $Q_c(x) = x^2 + c$. We know that, if $|x|$ is sufficiently large, then the orbit of x tends to infinity. Hence no basin of attraction extends to $\pm\infty$. Since 0 is the only critical point of Q_c and $SQ_c < 0$, it follows that Q_c has at most one attracting cycle.

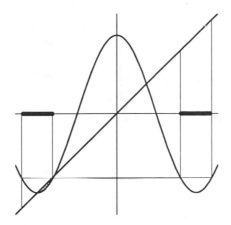

Fig. 12.2 The immediate basin of attraction of $-\pi$ for $C(x) = \pi \cos x$
is smaller than the full basin.

This is also the reason why we always use the critical point to plot the
orbit diagram for such functions as the quadratic or the logistic family. For
these families there is only one critical point. Hence, if these functions have
an attracting cycle, the critical point must "find" it.

There may be no critical points in the basin of attraction of attracting
cycles, as the following example shows. The above theorem guarantees that
such basins must extend to infinity, so there can be at most two such orbits.

Example. Consider the function $A_\lambda(x) = \lambda \arctan x$. Since

$$A'_\lambda(x) = \frac{\lambda}{1 + x^2},$$

A_λ has no critical points when $\lambda \neq 0$. If, however, $|\lambda| < 1$, then 0 is an
attracting fixed point. The immediate basin of attraction is the entire real
line. If $\lambda > 1$, then Figure 12.3 shows that A_λ has two attracting fixed points
and both basins extend to infinity. If $\lambda < -1$, A_λ has an attracting 2-cycle,
and again the immediate basin extends to infinity.

We conclude this chapter by providing a proof of this Theorem in the
simple case of a fixed point. For periodic points the proof is similar in spirit
but the details are more complicated.*

Proof: We will prove that the immediate basin of attraction of an attracting
fixed point p either contains a critical point or else extends to infinity.

---

*   Full details may be found in Devaney, page 72.

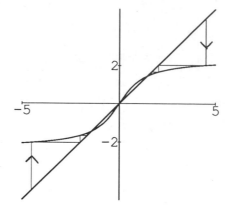

Fig. 12.3 Both immediate basins extend to infinity.
for $A_\lambda(x) = \lambda \arctan x$ when $\lambda > 1$.

The immediate basin of attraction of p must be an open interval, for otherwise, by continuity, we could extend the basin beyond the endpoints. So suppose the immediate basin of p is the interval (a, b). If either a or b are infinite we are done. So suppose both a and b are finite.

Since F maps the interval (a, b) to itself, it follows that F must preserve the endpoints of this interval. That is, $F(a)$ must be either a or b, and $F(b)$ must be a or b as well. Thus there are essentially four possibilities for the graph of F on the interval (a, b). They are:

1. $F(a) = a$, $F(b) = b$
2. $F(a) = b$, $F(b) = a$
3. $F(a) = a$, $F(b) = a$
4. $F(a) = b$, $F(b) = b$

See Figure 12.4 for a picture of these four possibilities. Note that, in cases 3 and 4, F must have a maximum or a minimum in (a, b), which is therefore attracted to p, so the theorem is true in these cases.

We deal first with case 1. Clearly, F can have no other fixed points besides p in (a, b). We claim that $F(x) > x$ in (a, p). To see this, note first that we cannot have $F(x) = x$ in (a, p), for this would give a second fixed point in (a, b). Also, if $F(x) < x$ for all x in (a, p), then graphical analysis shows that p is not an attracting fixed point. Consequently, we must have $F(x) > x$ in the interval (a, p). Similar arguments show that $F(x) < x$ in the interval (p, b).

The Mean Value Theorem asserts that there is a point c in (a, p) such that

$$F'(c) = \frac{F(a) - F(p)}{a - p} = \frac{a}{a - p} - \frac{p}{a - p} = 1.$$

Note that $c \neq p$ since $F'(p) < 1$. Similarly, there is a point d in (p, b) for which $F'(d) = 1$.

Thus, on the interval $[c, d]$, which contains p in its interior, we have

$$F'(c) = 1$$
$$F'(d) = 1$$
$$F'(p) < 1.$$

By the Schwarzian Min-Max Principle, F' cannot have a positive local minimum in $[c, d]$. Thus F' must become negative in $[c, d]$, so there is at least one point in the basin of attraction of p at which the derivative vanishes. This gives us a critical point in the basin.

To handle case 2, we consider $G(x) = F^2(x)$. The fixed point p is still attracting for G and (a, b) is the immediate basin of attraction of p under G. Moreover, $SG < 0$ by the Chain Rule for Schwarzian Derivatives. Since $G(a) = a$ and $G(b) = b$ the arguments of case 1 show that G must have a critical point $\hat{x}$ in (a, b). Since $G'(\hat{x}) = F'(F(\hat{x})) \cdot F'(\hat{x})$, it follows that one of $\hat{x}$ or $F(\hat{x})$ is a critical point of F in (a, b). This completes the proof.

As a final remark we note that the above arguments work if p is a neutral fixed point that attracts from one side. Such points must therefore have basins that extend to infinity or else must attract a critical point.

Exercises

1. Compute the Schwarzian derivative for the following functions and decide if $SF(x) < 0$ for all x.
   a. $F(x) = x^2$
   b. $F(x) = x^3$
   c. $F(x) = e^{3x}$
   d. $F(x) = \cos(x^2 + 1)$
   e. $F(x) = \arctan x$

2. Is it true that $S(F + G)(x) = SF(x) + SG(x)$? If so, prove it. If not, give a counterexample.

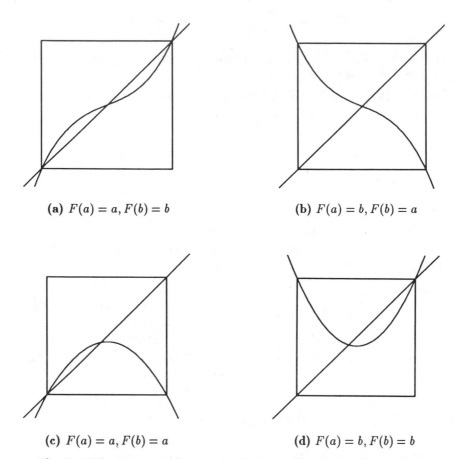

(a) $F(a) = a, F(b) = b$ **(b)** $F(a) = b, F(b) = a$

(c) $F(a) = a, F(b) = a$ **(d)** $F(a) = b, F(b) = b$

Fig. 12.4 The four possible cases for the immediate basin of attraction.

3. Is it true that $S(F \cdot G) = SF(x) \cdot G(x) + F(x) \cdot SG(x)$? If so, prove it. If not, give a counterexample.

4. Is it true that $S(cF)(x) = cSF(x)$ where c is a constant? If so, prove it. If not, give a counterexample.

5. Give an example of a function that has $SF(x) > 0$ for at least some x-values.

6. Prove that $S(1/x) = 0$ and $S(ax + b) = 0$. Conclude that $SF(x) = 0$ where

$$F(x) = \frac{1}{ax + b}.$$

7. Compute $SM(x)$ where

$$M(x) = \frac{ax + b}{cx + d}.$$

8. Let M be as in the previous exercise. Prove that $S(M \circ F) = SF$.

9. Give a formula for $S(F \circ G \circ H)(x)$ in terms of SF, SG, SH and the derivatives of these functions.

10. Compute the Schwarzian derivatives of each of the following functions:
     a. $F(x) = \tan x$
     b. $F(x) = \tan\left(\frac{1}{3x+4}\right)$
     c. $F(x) = \sin\left(e^{x^2+2}\right)$
     d. $F(x) = 1/\cos(x^2 - 2)$.

CHAPTER 13

Newton's Method

One of the basic applications of iteration is Newton's method—a classical algorithm for finding roots of a function. In this chapter we combine many of the ideas of previous chapters to give a detailed account of this topic.

13.1   Basic Properties

Consider the problem of trying to find the roots of a given function F, that is, solving the equation $F(x) = 0$. As is well known, this procedure can be carried out using algebraic methods such as factoring for only a few classes for functions such as low-degree polynomials. For other functions, we must resort to numerical methods. Among the simplest of these methods is Newton's method. This method (sometimes called the Newton-Raphson method) is predicated on the following idea. Suppose we try to "guess" a root x_0. Chances are that x_0 will not be a solution of the equation, so we use x_0 to produce a new point x_1, which will hopefully be closer to a root. The point x_1 is determined from x_0 as follows. Draw the tangent line to the graph of F at $(x_0, F(x_0))$. Unless we have made the unfortunate choice of x_0 so that $F'(x_0) = 0$, this tangent line is not horizontal and so meets the x-axis at a new point which we call x_1 (Fig. 13.1). This is our "new" choice for a root of F. We then iterate this procedure, with the hope that this process will eventually converge to a root. Sometimes this happens, as shown in Figure 13.2.

To find a formula for x_1 in terms of x_0, we first write down the equation for the tangent line to the graph of F at $(x_0, F(x_0))$. The slope of this line

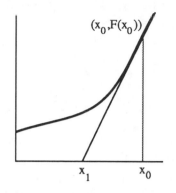

Fig. 13.1 Newton's method yields x_1 given x_0.

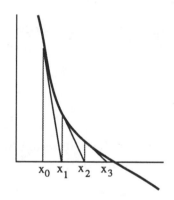

Fig. 13.2 Newton's method converges to a root of F.

is $F'(x_0)$. Hence the equation of the tangent line assumes the form

$$y = F'(x_0)x + B$$

where B is determined from the fact that $(x_0, F(x_0))$ lies on the line. Substituting this information for x and y in the above equation, we find

$$B = F(x_0) - F'(x_0)x_0.$$

Hence the tangent line is given by

$$y = F'(x_0)(x - x_0) + F(x_0).$$

Now x_1 is determined by setting $y = 0$ and solving for x. We find

$$x_1 = x_0 - \frac{F(x_0)}{F'(x_0)}.$$

This determines x_1 in terms of x_0. To apply Newton's method we iterate this procedure, determining in succession

$$x_2 = x_1 - \frac{F(x_1)}{F'(x_1)}$$

$$x_3 = x_2 - \frac{F(x_2)}{F'(x_2)}$$

and so forth. As shown in Figure 13.2, this sequence of points $x_0, x_1, x_2, \ldots$ sometimes converges to a root of F.

Thus we see that the question of finding roots of F may be recast as a problem of iteration, not of the function F, but rather of an associated function called the Newton iteration function.

Definition. Suppose F is a function. The *Newton iteration function* associated to F is the function

$$N(x) = x - \frac{F(x)}{F'(x)}.$$

Example. Consider $F(x) = x^2 - 1$. This function has two roots, at $x = \pm 1$. The associated Newton iteration function is

$$N(x) = x - \frac{x^2 - 1}{2x} = \frac{1}{2}\left(x + \frac{1}{x}\right).$$

The graph of N is displayed in Figure 13.3. Note that N has two fixed points, at the roots ± 1 of F. Note also that graphical analysis shows that the orbit of any nonzero point under N converges to one of these fixed points. Hence Newton's method succeeds in this case: if we make any initial guess $x_0 \neq 0$, the corresponding orbit of x_0 under N tends to one of the roots of F. Of course, one hardly needs Newton's method for this simple example, but this does illustrate how this method works.

In the above example, the roots of F appear as the fixed points of N. This is no accident, as we will see in a moment. But first we need a digression on the *multiplicity* of a root.

Definition. A root x_0 of the equation $F(x) = 0$ has *multiplicity* k if $F^{[k-1]}(x_0) = 0$ but $F^{[k]}(x_0) \neq 0$. Here $F^{[k]}(x_0)$ is the kth derivative of F and $F^{[0]} = F$.

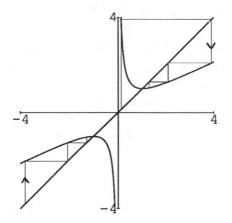

Fig. 13.3 Newton's method for $F(x) = x^2 - 1$.

For example, 0 is a root of multiplicity 2 for $F(x) = x^2 + x^3$ and of multiplicity 1 for $F(x) = x + x^3$. It can be shown that if x_0 is a root of multiplicity k for F, then $F(x)$ may be written in the form

$$F(x) = (x - x_0)^k G(x)$$

where G is a function that has no root at x_0. See exercise 11. If F is a polynomial, then the multiplicity of any root is always finite. However, there are examples of functions with roots of infinite multiplicity.*

Newton's Fixed Point Theorem. *Suppose F is a function and N is its associated Newton iteration function. Then x_0 is a root of F of multiplicity k if and only if x_0 is a fixed point of N. Moreover, such a fixed point is always attracting.*

Proof. Suppose for the moment that $F(x_0) = 0$ but $F'(x_0) \neq 0$, that is, the root has multiplicity 1. Then we have $N(x_0) = x_0$ so x_0 is a fixed point of N. Conversely, if $N(x_0) = x_0$ we must also have $F(x_0) = 0$.

To see that x_0 is an attracting fixed point, we use the quotient rule to compute

$$N'(x_0) = \frac{F(x_0)F''(x_0)}{(F'(x_0))^2}.$$

---

*   Consider $G(x) = \exp(-1/x^2)$ if $x \neq 0$ and set $G(0) = 0$. It can be shown that the kth derivative of G vanishes at 0 for all k. So 0 is a root of G with infinite multiplicity. See exercise 10.

Again assuming $F'(x_0) \neq 0$, we see that $N'(x_0) = 0$ so that x_0 is indeed an attracting fixed point.

This proves the theorem subject to the special assumption that $F'(x_0) \neq 0$. If $F'(x_0) = 0$, we have to work harder. Let's suppose that the root has multiplicity $k > 1$ so that the $(k-1)$th derivative of F vanishes at x_0 but the kth does not. Thus we may write

$$F(x) = (x - x_0)^k G(x)$$

where G is a function that satisfies $G(x_0) \neq 0$. Then we have

$$F'(x) = k(x - x_0)^{k-1} G(x) + (x - x_0)^k G'(x)$$

$$F''(x) = k(k-1)(x - x_0)^{k-2} G(x) + 2k(x - x_0)^{k-1} G'(x) + (x - x_0)^k G''(x).$$

Therefore, after some cancellation, we have

$$N(x) = x - \frac{(x - x_0) G(x)}{k G(x) + (x - x_0) G'(x)}.$$

Hence $N(x_0) = x_0$, showing that roots of F correspond to fixed points of N in this case as well. Finally we compute

$$N'(x) = \frac{k(k-1)(G(x))^2 + 2k(x - x_0)G(x)G'(x) + (x - x_0)^2 G(x)G''(x)}{k^2(G(x))^2 + 2k(x - x_0)G(x)G'(x) + (x - x_0)^2(G'(x))^2}$$

where we have factored out $(x - x_0)^{2k-2}$ from both numerator and denominator. Now $G(x_0) \neq 0$, so

$$N'(x_0) = \frac{k-1}{k} < 1.$$

Thus, we again see that x_0 is an attracting fixed point for N. This completes the proof.

Example. Consider $F(x) = x^2(x - 1)$ so $N(x) = x - \frac{x^2 - x}{3x - 2}$. Note that 0 and 1 are roots. We compute $F'(1) = 1$, $F'(0) = 0$, but $F''(0) = -2$. Hence $N'(1) = 0$ and $N'(0) = 1/2$. The graphs of F and N are depicted in Figure 13.4.

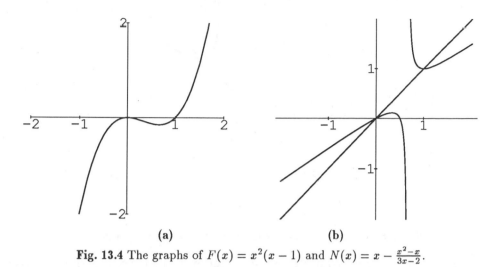

(a)   (b)

Fig. 13.4 The graphs of $F(x) = x^2(x-1)$ and $N(x) = x - \frac{x^2-x}{3x-2}$.

13.2   Convergence and Nonconvergence

Newton's method, unfortunately, does not always converge. That is, a given initial guess need not lead to an orbit that tends to one of the attracting fixed points of N which are, by the Newton Fixed Point Theorem, the roots of F.

One problem that arises occurs when the function F is not differentiable at the root.

Example. Consider $F(x) = x^{1/3}$. This function is not differentiable at the root $x = 0$. Note that $N(x) = -2x$, which has a repelling fixed point at 0. Moreover, all other orbits tend to infinity. Hence we may have no convergence if there is no differentiability.

Assuming that F is differentiable may still yield problems. For example, as we know, it is entirely possible for a dynamical system such as N to have periodic orbits, which therefore do not tend to fixed points.

Example. Consider $F(x) = x^3 - 5x$. We compute

$$N(x) = x - \frac{x^3 - 5x}{3x^2 - 5}.$$

The graph of N is shown in Figure 13.5. Suppose we make the natural but unfortunate initial guess $x_0 = 1$. Then $N(1) = -1$ and $N(-1) = 1$ so that ± 1 lie on a 2 cycle. Hence this initial guess does not lead to convergence.

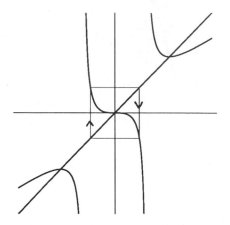

Fig. 13.5 Newton's method for $F(x) = x^3 - 5x$ showing a 2-cycle.

Note that, in this example, most other initial guesses lead to convergence. However, it may happen that intervals of initial guesses lead to non-convergence.

Example. Consider $F(x) = (x^2 - 1)(x^2 + A)$ for real values of A. We compute

$$N(x) = \frac{3x^4 + (A - 1)x^2 + A}{4x^3 + 2(A - 1)x}.$$

Note that $N(-x) = -N(x)$, so N is an odd function. As we saw earlier, N has critical points at the roots of F and also at the points where $F''(x)$ vanishes. Hence we find that the points

$$c_\pm = \pm\sqrt{\frac{1 - A}{6}}$$

are critical points for N. In Figure 13.6, the graph of N is shown in the special case where $A = 0.197017\ldots$. Note that, in this case, $c_\pm$ both lie on a cycle, which therefore must be attracting. Hence there is an open interval about each of these points whose orbits converge to the 2-cycle. Thus, Newton's method fails to converge to roots for these initial guesses.

The exact value of A for which $c_\pm$ lie on a 2-cycle is $(29 - \sqrt{720})/11$, as we ask you to show in exercise 8.

However, it may happen that no initial guesses lead to convergence. This happens, for example, if F has no roots.

Example. Consider $F(x) = x^2 + 1$. Obviously, F has no roots on the real line. But let's see what happens when we attempt to use Newton's method

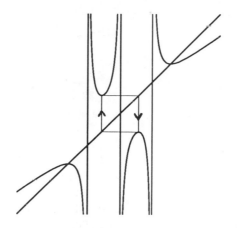

Fig. 13.6 Newton's method with an attracting 2-cycle.

anyway. The Newton iteration function is $N(x) = \frac{1}{2}(x - \frac{1}{x})$. Graphical analysis of N is shown in Figure 13.7. Typical initial conditions seem to lead to orbits that wander around the real line aimlessly. So we get no convergence to a root, as we expected.

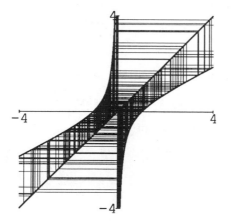

Fig. 13.7 Chaos in Newton's method for $F(x) = x^2 + 1$.

In fact, N is chaotic in the sense of Chapter 10. Recall the doubling function

$$D(x) = \begin{cases} 2x & 0 \leq x < 1/2 \\ 2x - 1 & 1/2 \leq x < 1. \end{cases}$$

As we saw in Chapter 10 (see exercise 20), this function behaves quite chaot-

n	$x(x+1)$	$x^2(x+1)$
0	100	100
1	49.751...	66.566...
2	24.628...	44.260...
3	12.069...	29.397...
4	5.794...	19.489 ...
5	2.667...	12.885...
6	1.123...	8.484...
7	0.388...	5.553...
8	0.084...	3.603 ...
9	0.006...	2.308...
10	0.000...	1.452...
11	0.000...	0.892...
12	0.000...	0.531...
13	0.000...	0.304...
14	0.000...	0.168...
15	0.000...	0.089...
16	0.000...	0.046...
17	0.000...	0.023...
18	0.000...	0.012...
19	0.000...	0.006...
20	0.000...	0.003...

Table 13.1 Orbits for Newton's method applied to $x(x+1)$ and $x^2(x+1)$.

ically. Define $C\colon [0,1) \to \mathbf{R}$ by $C(x) = \cot(\pi x)$. Then we have

$$C \circ D(x) = \cot(\pi \cdot D(x))$$
$$= \cot(2\pi x)$$
$$= \frac{\cos^2(\pi x) - \sin^2(\pi x)}{2\sin(\pi x)\cos(\pi x)}$$
$$= \frac{1}{2}\left(\cot(\pi x) - \frac{1}{\cot(\pi x)}\right)$$
$$= N \circ C(x).$$

Therefore, the Newton iteration function for $F(x) = x^2 + 1$ is conjugate to our friend the doubling function. Consequently, N is chaotic on the entire real line.

One of the reasons for the importance of Newton's method is the speed with which it converges. Recall that if x_0 is a root of F for which $F'(x_0) \neq 0$, then x_0 is an attracting fixed point for N with $N'(x_0) = 0$. Fixed points whose derivative is 0 are called *superattracting* fixed points because nearby orbits are attracted to them very quickly. In Table 13.1 we have listed the orbit of 100 for the Newton iteration function associated to both $x(x + 1)$ and $x^2(x + 1)$. In the first case, the orbit converges rapidly; in the second, it converges to 0, but not nearly as quickly. The reason is that Newton's method for $x^2 + x$ has superattracting fixed points whereas, at the root 0 for $x^2(x + 1)$, we have $N'(0) = 1/2$.

There are many other methods for finding roots of functions, and many of them involve iteration. Most methods suffer from the deficiency that they do not always work. Some methods, when they do converge, do so more rapidly than others. The trade-off is always how often an algorithm works (its efficiency) versus how quickly it converges (its speed). A full study of the speed and efficiency of algorithms to find roots is one of the topics of the field of mathematics known as numerical analysis.

Exercises

1. Use graphical analysis to describe completely all orbits of the associated Newton iteration function for F when

    a. $F(x) = 4 - 2x$

    b. $F(x) = x^2 - 2x$

    c. $F(x) = x^{2/3}$

    d. $F(x) = x^4 + x^2$

    e. $F(x) = 1/x$

    f. $F(x) = \frac{1}{x} - 1$

    g. $F(x) = x/\sqrt{1 + x^2}$

    h. $F(x) = xe^x$.

2. What happens when Newton's method is applied to $F(x) = \sqrt{x}$?

3. Find all fixed points for the associated Newton iteration function for $F(x) = x/(x - 1)^n$ when $n = 1, 2, 3 \ldots$. Which are attracting and which are repelling?

4. Consider the Newton iteration function for $F(x) = \sec x$. What are the fixed points for N? Does this contradict the Newton Fixed Point Theorem? Why or why not?

5. Suppose $P(x)$ and $Q(x)$ are polynomials and let $F(x) = P(x)/Q(x)$. What can be said about the fixed points of the associated Newton function for F? Which fixed points are attracting and which are repelling?

6. A bifurcation. Consider the family of functions $F_\mu(x) = x^2 + \mu$. Clearly, F_μ has two roots when $\mu < 0$, one root when $\mu = 0$, and no real roots when $\mu > 0$. Your goal in these exercises is to investigate how the dynamics of the associated Newton iteration function changes as μ changes.

a. Sketch the graphs of the associated Newton iteration function N_μ in the three cases $\mu < 0$, $\mu = 0$, and $\mu > 0$.

b. Use graphical analysis to explain the dynamics of N_μ when $\mu < 0$ and $\mu = 0$.

c. Prove that, if $\mu > 0$, the Newton iteration function for F_1 is conjugate to the Newton iteration function for F_μ via the conjugacy $H(x) = \sqrt{\mu}x$. Conclude that the Newton iteration function is chaotic for $\mu > 0$.

d. Find an analogous conjugacy when $\mu < 0$.

7. A more complicated bifurcation. Consider the family of functions given by $G_\mu(x) = x^2(x-1) + \mu$.

a. Sketch the graphs of the associated Newton iteration function N_μ in the three cases $\mu > 0$, $\mu = 0$, and $\mu < 0$.

b. Use graphical analysis to discuss the fate of all orbits in case $\mu = 0$.

c. Show that N_μ has exactly one critical point that is not fixed for all but one μ-value.

d. What is the fate of the orbit of this critical point when $\mu > 0$? Describe using graphical analysis.

e. Now consider the case $\mu < 0$. In an essay, describe possible fates of the orbit of this "free" critical point. Can you find μ-values for which this critical point is periodic? (This can be done experimentally with our iteration programs.) If this critical point is periodic (not fixed) for a given μ-value, what can you conclude about the convergence of Newton's method?

8. Consider the function $G(x) = x^4 - x^2 - 11/36$.

a. Compute the inflection points of G. Show that they are critical points for the associated Newton function.

b. Prove that these two points lie on a 2-cycle.

c. What can you say about the convergence of Newton's method for this function?

9. Use calculus to sketch the graph of the Newton iteration for $F(x) = x(x^2 + 1)$. For which x-values does this iteration converge to a root?

10. Prove that the critical points for the Newton iteration function associated to

$$F(x) = (x^2 - 1)(x^2 + A)$$

lie on a 2-cycle when $A = (29 - \sqrt{720})/11$.

11. Prove that the equation $F(x) = 0$ has a root of multiplicity k at x_0 if and only if $F(x)$ may be written in the form

$$F(x) = (x - x_0)^k G(x)$$

where G does not vanish at x_0. *Hint:* Use the Taylor expansion of F about x_0.

12. Let $G(x) = \exp(-1/x^2)$ if $x \neq 0$ and set $G(0) = 0$. Compute the Newton iteration function N for G. What can be said about the fixed point of N? Why does this occur?

CHAPTER 14

Fractals

Technically, there is no connection between the fields of dynamical systems and fractal geometry. Dynamics is the study of objects in motion such as iterative processes; fractals are geometric objects that are static images. However, it has become apparent in recent years that most chaotic regions for dynamical systems are fractals. Hence, in order to understand chaotic behavior completely, we must pause to understand the geometric structure of fractals. This is the topic of this chapter. We have seen the prototypical fractal, the Cantor set, when we discussed the quadratic mapping in Chapter 7. We will see many more such objects when we discuss complex dynamics in Chapters 16–18.

14.1   The Chaos Game

Before we begin the study of the geometry of fractal sets, we pause to show how these sets arise as the "chaotic regime" for a dynamical system. Consider the following process which has been called* the "Chaos Game." Begin with three points A, B, and C in the plane forming the vertices of an equilateral triangle. Now choose any point p_0 in the plane as the initial seed for the iteration. Note that p_0 need not be chosen inside the triangle whose vertices are A, B, and C.

The next point on the orbit of p_0 is determined by randomly choosing one of A, B, or C and then moving p_0 halfway toward the selected vertex.

---

*   See M. Barnsley, *Fractals Everywhere,* New York: Academic Press, 1988.

That is, p_1 is the midpoint of the line between p_0 and the chosen vertex.*

To continue, we again choose one of A, B, or C randomly and then let p_2 be the midpoint of the line segment between p_1 and the chosen vertex. In general, p_{n+1} is obtained from p_n similarly.

The sequence of points $p_0, p_1, p_2 \ldots$ is, as usual, called the *orbit* of p_0. As with any dynamical system, the question is what is the fate of the orbit of p_0 under this iteration? It appears that the answer to this question would be heavily dependent upon which sequence of vertices we choose. For example, if we always choose A as the vertex (hardly a random choice!), then it is clear that the p_i simply converge to A. If we always choose vertex B, then the p_i converge to B.

One of the amazing surprises in this game is the fact that, as long as we make random choices of A, B, and C, the fate of the orbit of p_0 does not depend on these choices or on the choice of p_0. If we choose any initial seed whatsoever, the orbit tends (with probability 1) to the same set. In Figure 14.1 we have displayed the results of randomly iterating this procedure 20,000 times. (We have not pictured the first 50 points on this orbit so that only the eventual behavior is shown). Note the intricate shape of the resulting orbit. This figure, called the Sierpinski triangle or gasket, is a classical example of a *fractal*.

Fig. 14.1 The Sierpinski triangle.

---

*   A nice way of describing this is to let A represent 1 and 2 on a die, B represent 3 and 4, and C represent 5 and 6. Rolling the die then provides the mechanism for determining the vertex toward which we move at each step.

Benoit B. Mandelbrot

In 1975, *Benoit Mandelbrot* (1924–) observed that geometric objects like the Cantor set and the Sierpinski triangle were not mathematical pathologies. Rather, these complicated sets provided a geometry that is in many ways more natural than classical Euclidean geometry for describing intricate objects in nature such as coastlines and snowflakes. Thus was born fractal geometry. As we will see in the remainder of this book, fractals also form the typical set on which a dynamical system behaves chaotically. Mandelbrot also paved the way for the resurgence of interest in complex dynamics with his discovery in 1980 of the Mandelbrot set, our topic in Chapter 17. He is currently Abraham Robinson Professor of Mathematical Sciences at Yale University and IBM Fellow at the IBM Thomas J. Watson Research Center.

There are a number of different definitions of fractals that are currently in use. We prefer the following definition:

Definition. A *fractal* is a subset of $\mathbf{R}^n$ which is self-similar and whose fractal dimension exceeds its topological dimension.

Obviously, some terms in this definition need explanation. We will see that the fractal dimension of a set need not be an integer, so this will necessitate careful explanation as well. We will elaborate on these concepts after first giving three classical examples of fractal sets.

We must emphasize that there are many other possible definitions of a fractal set. In particular, there are many other possible notions of the dimension of a set, including Hausdorff dimension, capacity dimension, correlation dimension, and others. We will consider here only the most elementary of these notions—fractal dimension.

14.2   The Cantor Set Revisited

We begin the study of the geometry of fractal sets by considering three important examples. The first is an old friend, the Cantor middle-thirds set. We have seen already that sets like this occur naturally as the chaotic regime for the quadratic family $Q_c(x) = x^2 + c$ when $c < -2$.

Recall that the Cantor middle-thirds set C is obtained by successively re-moving open middle thirds of intervals from the unit interval. The procedure is depicted in Figure 14.2.

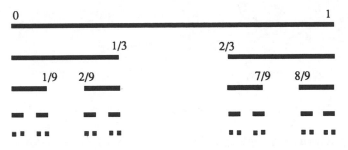

Fig. 14.2 Construction of C, the Cantor middle-thirds set.

One of the most important properties of a fractal is known as *self-similarity*. Roughly speaking, self-similarity means if we examine small por-tions of the set under a microscope, the image we see resembles our original set. For example, look closely at C. Note that C may be decomposed into two distinct subsets, the portion of C in $[0, 1/3]$ and the portion in $[2/3, 1]$. If we examine each of these pieces, we see that they resemble the original Cantor set C. Indeed, each is obtained by removing middle-thirds of inter-vals. The only difference is the original interval is smaller by a factor of $1/3$. Thus, if we magnify each of these portions of C by a factor of 3, we obtain the original set.

More precisely, to magnify these portions of C, we use an *affine trans-formation*. Let $L(x) = 3x$. If we apply L to the portion of C in $[0, 1/3]$, we see that L maps this portion onto the entire Cantor set. Indeed, L maps $[1/9, 2/9]$ to $[1/3, 2/3]$, $[1/27, 2/27]$ to $[1/9, 2/9]$, and so forth (Fig. 14.3). Each of the gaps in the portion of C in $[0, 1/3]$ is taken by L to a gap in C. That is, the "microscope" we use to magnify $C \cap [0, 1/3]$ is just the affine transformation $L(x) = 3x$.

To magnify the other half of C, namely $C \cap [2/3, 1]$, we use another affine transformation, $R(x) = 3x - 2$. Note that $R(2/3) = 0$ and $R(1) = 1$ so R takes $[2/3, 1]$ linearly onto $[0, 1]$. As with L, R takes gaps in $C \cap [2/3, 1]$ to gaps in C, so R again magnifies a small portion of C to give the entire set.

Using more powerful "microscopes," we may magnify arbitrarily small portions of C to give the entire set. For example, the portion of C in $[0, 1/3]$ itself decomposes into two self-similar pieces: one in $[0, 1/9]$ and one in $[2/9, 1/3]$. We may magnify the left portion via $L_2(x) = 9x$ to yield

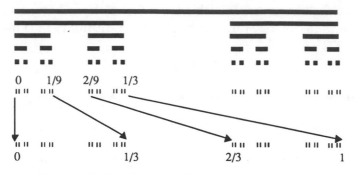

Fig. 14.3 Self-similarity of the Cantor middle-thirds set

C and the right portion via $R_2(x) = 9x - 2$. Note that R_2 maps $[2/9, 1/3]$ onto $[0, 1]$ linearly as required.

Note also that, at the nth stage of the construction of C, we have 2^n small copies of C, each of which may be magnified by a factor of 3^n to yield the entire Cantor set.

14.3 The Sierpinski Triangle

Now let's consider another fractal set, the Sierpinski triangle (sometimes called the Sierpinski gasket), which we encountered while playing the chaos game. Like the Cantor middle-thirds set, this object may also be obtained via an infinite sequence of "removals." Begin with the equilateral triangle shown in Figure 14.4. Then remove from the middle a triangle whose dimensions are exactly half that of the original triangle. This leaves three smaller equilateral triangles, each of which has dimensions one-half the dimensions of the original triangle. Now continue this process. Remove the middle portions of each of the remaining triangles, leaving nine equilateral triangles, and so forth. The resulting image after carrying this procedure to the limit is denoted T and is called the Sierpinski triangle. This is essentially the same set displayed in Figure 14.1. Note how the same image may be obtained in two remarkably different ways: via the deterministic process of removing triangles as above and also via the probabilistic methods of the chaos game.

Like the Cantor middle-thirds set, the Sierpinski triangle is also self-similar. This time, however, the magnification factor is 2. For example, after removing the middle equilateral triangle in the first step of this construction, we are left with three smaller copies of T, each of whose dimensions are one-half the dimension of the entire triangle. At the nth stage of this

Fig. 14.4 Constructing the Sierpinski triangle.

construction, we have 3^n Sierpinski triangles, each of which may be magnified by a factor of 2^n to yield the entire set.

There are many fractals that may be constructed via variations on this theme of infinite removals. For example, we may construct a similar set by beginning with a right triangle, as in Figure 14.5. Another fractal, the "box" fractal, is obtained by successively removing squares whose sides are one-third as long as their predecessors, as shown in Figures 14.6 and 14.7.

Fig. 14.5 Construction of another Sierpinski triangle.

Note that both of these sets are connected. If we remove larger triangles or squares at each stage, then the resulting set is totally disconnected as in

Fig. 14.6 Construction of the box fractal.

Fig. 14.7 The box fractal.

the case of the Cantor middle-thirds set.

14.4   The Koch Snowflake

Unlike the Sierpinski triangle, the Koch snowflake is generated by an infinite succession of additions. This time we begin with the boundary of an equilateral triangle with sides of length 1. The first step in the process is to remove the middle third of each side of the triangle, just as we did in the construction of the Cantor set. This time, however, we replace each of these pieces with two pieces of equal length, giving the star-shaped region depicted in Figure 14.8. This new figure has twelve sides, each of length $\frac{1}{3}$. Now we iterate this process. From each of these sides we remove the middle third and replace it with a triangular "bulge" made of two pieces of length 1/9. The result is also shown in Figure 14.8.

We continue this process over and over. The ultimate result is a curve that is infinitely wiggly—there are no straight lines in it whatsoever. This object is called the Koch snowflake.

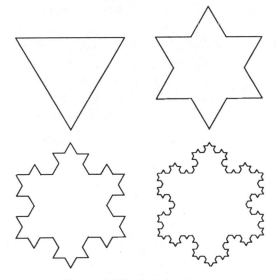

Fig. 14.8 The first four stages
in the construction of the Koch snowflake.

Clearly, there are pieces of the snowflake that are self-similar. Suppose we look at one side of the snowflake. What we see is called the *Koch curve* and is depicted in Figure 14.9. If we examine one third of this edge and magnify this portion by a factor of 3, we again see the same figure. Note that there are exactly four pieces of the snowflake's edge that, when magnified by a factor of 3 (and possibly rotated), yield the entire edge of the snowflake.

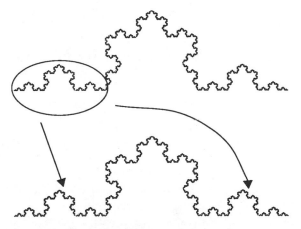

Fig. 14.9 Magnification of the Koch curve.

At each stage of the construction of the Koch curve, magnification by a factor of 3 yields the previous image. As before, this means that the ultimate figure is self-similar.

This fractal snowflake has an amazing geometric property: It has finite area but its perimeter is infinite! This means that we can paint the inside of the Koch snowflake, but we can never wrap a length of string around its boundary! This is quite a contrast to the usual shapes encountered in geometry such as squares and circles, which have finite area *and* perimeter.

To see why this is true, let $N_0, N_1, N_2, \ldots$ denote the number of sides of the snowflake at the corresponding stage of the construction. We find

$$N_0 = 3$$
$$N_1 = 4 \cdot 3 = 12$$
$$N_2 = 4 \cdot 12 = 4^2 \cdot 3$$
$$\vdots$$
$$N_k = 4N_{k-1} = 4^k \cdot 3.$$

These numbers get large very quickly. For example, N_8 gives 196,608 little sides.

Now let's compute the perimeter. Let L_k be the length of one segment of the perimeter after the kth stage. At the beginning, each side has length 1; after the first addition, each side has length $1/3$; after the second, each side has length $1/3^2$, and so forth. We find that, at the kth stage,

$$L_k = \frac{1}{3^k}.$$

Now let P_k be the perimeter of the figure at the kth stage. Clearly, $P_k = N_k \cdot L_k$, so we have

$$P_k = N_k \cdot L_k = 4^k \cdot 3 \cdot \frac{1}{3^k}$$
$$= \left(\frac{4}{3}\right)^k \cdot 3.$$

Hence $P_k \to \infty$ as $k \to \infty$.

The area contained within the snowflake is more difficult to compute, but you can easily check using plane geometry that the snowflake is contained within a square in the plane whose sides have length $2\sqrt{3}/3$. Therefore this area is certainly less than $4/3$.

14.5   Topological Dimension

As we have seen in the previous three sections, one distinguishing feature of a fractal is self-similarity. Each of the Cantor set, the Sierpinski triangle, and the Koch snowflake shares this property. But so do lines and squares and cubes: these familiar figures from Euclidean geometry are all self-similar. For example, a line segment may obviously be subdivided into N smaller subintervals of equal length each of which may be magnified by a factor of N to yield the original line segment. So what distinguishes complicated sets like Cantor sets and Sierpinski triangles from lines and planes? The answer is their *dimension*.

There is no question that a line is one-dimensional. Similarly, a square is two-dimensional, and a cube is three-dimensional. Naively speaking, the reasons for this are obvious. There is only one "linearly independent" direction to move along a line (backwards and forwards), two directions in a square (length and width), and three directions in a cube (length, width, and height). So these figures have dimensions 1, 2, and 3 respectively. But what is the dimension of the Sierpinski triangle? Sometimes it seems that the triangle has dimension 1—after all, we have removed all of the planar regions. On the other hand, the Sierpinski triangle is clearly much more complicated than a typical one-dimensional object like a line or a curve. Also, we can move in the Sierpinski triangle in many directions—not just one—but clearly we cannot move in every planar direction. So what then is the dimension of the Sierpinski triangle? At times this figure seems one-dimensional; at other times it seems two-dimensional. A nice experiment to perform is to take a vote among your classmates as to what is the dimension of this figure: one or two. If you average the results you'll see that the answer is somewhere in between 1 and 2. Many people say that the Sierpinski triangle is two-dimensional, but a sizeable minority votes for one-dimensional. The votes in my classes often average 1.6, very nearly the accurate "fractal" dimension. The point is that these fractal images do not fit neatly into any of our preconceived notions of dimension, so we must first re-evaluate what dimension means before assigning a dimension to a set.

One of the crudest measurements of dimension is the notion of topological dimension. This dimension agrees with our naive expectation that a set should have an integer dimension. We define the topological dimension inductively.

Definition. A set S has *topological dimension* 0 if every point has arbitrarily small neighborhoods whose boundaries do not intersect the set.

For example, a scatter of isolated points has topological dimension 0, since each point may be surrounded by arbitrarily small neighborhoods whose boundaries are disjoint from the set (see Figure 14.10). In particular, the Cantor middle-thirds set has topological dimension 0 since any two points in C are separated by at least one gap in the Cantor set.

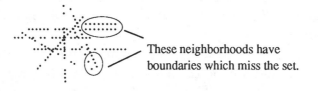

These neighborhoods have
boundaries which miss the set.

Fig. 14.10 A set with topological dimension 0.

To define topological dimension k, we use induction.

Definition. A set S has *topological dimension* k if each point in S has arbitrarily small neighborhoods whose boundaries meet S in a set of dimension $k - 1$, and k is the least nonnegative integer for which this holds.

For example, a line segment in the plane has topological dimension 1 since small disks in the plane have boundaries that meet the line in one or two points. Similarly, a planar region has topological dimension 2 because points in the set have arbitrarily small neighborhoods whose boundaries are one-dimensional, as depicted in Figure 14.11.

What about the Sierpinski triangle? As shown in Figure 14.12, we may surround points in T with arbitrarily small ovals that meet T in only finitely many points. Hence T has topological dimension 1.

14.6   Fractal Dimension

A more sensitive notion of dimension is provided by *fractal dimension* (sometimes called *similarity dimension*). Not all sets have a well-defined fractal dimension: we will consider only those sets that do, namely, those that are affine self-similar.

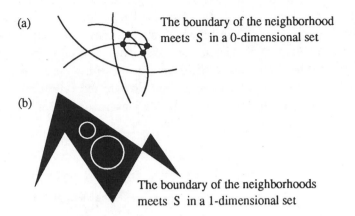

(a)

The boundary of the neighborhood meets S in a 0-dimensional set

(b)

The boundary of the neighborhoods meets S in a 1-dimensional set

Fig. 14.11 A set with topological dimension **(a)** 1, **(b)** 2.

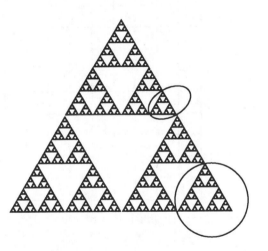

Fig. 14.12 Each neighborhood meets the Sierpinski Triangle in a finite set of points.

Definition. A set S is called *affine self-similar* if S can be subdivided into k congruent subsets, each of which may be magnified by a constant factor M to yield the whole set S.

Note that all of the sets considered earlier in this chapter are affine self-similar. Also, the line, the plane, and the cube are affine self-similar. We may now use this fact to provide a different notion of dimension, for one way to realize that these objects have different dimensions is to do the following. A line is a very self-similar object: It may be decomposed into $n = n^1$ little "bite-size" pieces, each of which is exactly $\frac{1}{n}$ the size of the original line and each of which, when magnified by a factor of n, looks exactly like the whole line (Fig. 14.13). On the other hand, if we decompose a square into pieces that are $\frac{1}{n}$ the size of the original square, then we find we need n^2 such pieces to reassemble the square. Similarly, a cube may be decomposed into n^3 pieces, each $\frac{1}{n}$ the size of the original. So the exponent in each of these cases distinguishes the dimension of the object in question. This exponent is the *fractal dimension*.

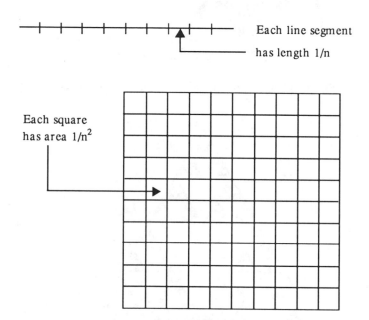

Fig. 14.13 Calculating the dimensions of a line and a square.

In these simple cases, it is trivial to read the exponent and find the dimension. For more general affine self-similar figures, this is not always as easy. However, note that we may also calculate this exponent as follows.

Definition. Suppose the affine self-similar set S may be subdivided into k congruent pieces, each of which may be magnified by a factor of M to yield the whole set S. Then the *fractal dimension* D of S is

$$D = \frac{\log(k)}{\log(M)} = \frac{\log(\text{number of pieces})}{\log(\text{magnification factor})}.$$

This definition agrees with the exponents we can already compute, since we have for a line:

$$\log(\text{number of pieces}) = \log(n^1) = 1\log n.$$

For a square,

$$\log(\text{number of pieces}) = \log(n^2) = 2\log n,$$

and for a cube,

$$\log(\text{number of pieces}) = \log(n^3) = 3\log n.$$

Since in each case the magnification factor is n, we find, for a line,

$$D = \frac{\log n^1}{\log n} = 1,$$

for a square,

$$D = \frac{\log n^2}{\log n} = \frac{2\log n}{\log n} = 2,$$

and for a cube,

$$D = \frac{\log n^3}{\log n} = \frac{3\log n}{\log n} = 3.$$

For the Sierpinski triangle, recall that we may subdivide this figure into three congruent triangles, each of which may be magnified by a factor of 2 to yield the whole figure. Thus we have

$$D = \frac{\log(\text{number of triangles})}{\log(\text{magnification})} = \frac{\log 3}{\log 2} = 1.584\ldots,$$

which is by no means an integer! Let's try this again. The Sierpinski triangle may also be constructed by assembling nine smaller pieces as we described in Section 14.3. Each of these smaller triangles is exactly one-fourth the size of the original figure. Hence

$$D = \frac{\log 9}{\log 4} = \frac{\log 3^2}{\log 2^2} = \frac{2\log 3}{2\log 2} = \frac{\log 3}{\log 2},$$

and we get the same answer.

To calculate the dimension of the Koch curve, we recall that each side of the original triangle is decomposed into four smaller pieces with a magnification factor of 3. Therefore,

$$D = \frac{\log 4}{\log 3} = 1.261\ldots.$$

We use the sides of the snowflake because no piece of the snowflake may be magnified to look like the whole object; pieces of the sides are self-similar, however. If we proceed to the second stage of the construction, there are then 16 sides, but the magnification factor is 3^2. Again,

$$D = \frac{\log 4^2}{\log 3^2} = \frac{2\log 4}{2\log 3} = 1.261\ldots.$$

Notice that the dimension of the curve is somewhat smaller than that of the Sierpinski triangle. This agrees with what our eyes are telling us (if eyes could speak). The triangle looks larger, more two-dimensional, than the Koch curve, and so it should have a larger dimension.

Finally, for the Cantor set, the number of intervals at each stage of the construction is 2^n, but the magnification factor is 3^n. So,

$$D = \frac{\log 2^n}{\log 3^n} = \frac{n\log 2}{n\log 3} = 0.6309\ldots.$$

Remark. The dimensions computed in this section were easy to compute because the magnification factor always increased at the same rate as the number of pieces in the figure. For many fractals, this is not the case. For example, the Julia sets that we will encounter in the next few chapters usually have fractional dimension, but this dimension is very difficult to compute. Indeed, for many Julia sets, the exact dimension is unknown.

14.7   Iterated Function Systems

We now return to the chaos game to show that a great many fractals may be obtained by variations on this theme. We use vector notation to denote points in the plane. Let

$$p_0 = \begin{pmatrix} x_0 \\ y_0 \end{pmatrix}$$

be a point in the plane and suppose $0 < \beta < 1$. The function

$$A \begin{pmatrix} x \\ y \end{pmatrix} = \beta \cdot \begin{pmatrix} x - x_0 \\ y - y_0 \end{pmatrix} + \begin{pmatrix} x_0 \\ y_0 \end{pmatrix}$$

has a fixed point at p_0 since $A(p_0) = p_0$. Since $\beta < 1$, it follows that A moves any point in the plane closer to p_0. Indeed, if

$$p = \begin{pmatrix} x \\ y \end{pmatrix},$$

then $A(p) - A(p_0) = \beta(p - p_0)$ so the distance between p and p_0 is contracted by a factor of β. The function A is an example of a *linear contraction* of the plane.

Note that the orbit of any point p in the plane converges to p_0 under iteration of A as shown in Figure 14.14.

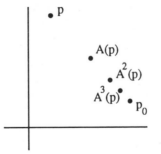

Fig. 14.14 Orbits under iteration of A converge to the fixed point p_0 if $0 < \beta < 1$.

The chaos game was played by randomly iterating three different contractions. The fixed points for these functions were the vertices of the original triangle and the contraction ratio was in each case $\beta = 1/2$. Note that, as we saw in Section 14.3, $2 = 1/\beta$ was the magnification factor at the first stage of the removals which generated the Sierpinski triangle. We may therefore construct different fractals by choosing different β-values and various contractions.

Definition. Let $0 < \beta < 1$. Let $p_1, \ldots, p_n$ be points in the plane. Let $A_i(p) = \beta(p - p_i) + p_i$ for each $i = 1, \ldots, n$. The collection of functions $\{A_1, \ldots, A_n\}$ is called an *iterated function system*.

To produce a fractal, we choose an arbitrary initial point in the plane and compute its orbit under random iteration of the A_i. It can be proved* that this orbit converges with probability 1 to a specific subset of the plane.

Definition. Suppose $\{A_1, \ldots, A_n\}$ is an iterated function system. The set of points to which an arbitrary orbit in the plane converges is the *attractor* for the system.

Example. The Cantor middle-thirds set may be obtained as the attractor of an iterated function system by setting

$$A_0 \begin{pmatrix} x \\ y \end{pmatrix} = \frac{1}{3} \begin{pmatrix} x \\ y \end{pmatrix}$$

$$A_1 \begin{pmatrix} x \\ y \end{pmatrix} = \frac{1}{3} \begin{pmatrix} x - 1 \\ y \end{pmatrix} + \begin{pmatrix} 1 \\ 0 \end{pmatrix}.$$

The contraction factor here is 1/3, and the fixed points are located at 0 and 1 along the x-axis.

Let's see that any orbit of the iterated function system tends to the Cantor set. Suppose p_n is the nth point on the orbit with

$$p_n = \begin{pmatrix} x_n \\ y_n \end{pmatrix};$$

then $y_n = y_0/3^n$, since $y_{k+1} = y_k/3$ no matter which of the two contractions we apply. Hence the orbit of any point tends toward the x-axis at a geometric rate.

What happens to the x-coordinates of points under this iteration? This depends upon which sequence of iterations we perform. We may describe this sequence of iterations by means of a sequence of 0's and 1's given by $(s_1 s_2 s_3 \ldots)$ where each s_j is either 0 or 1. We agree that if $s_j = 0$, the jth iteration performed is A_0; if $s_j = 1$, we apply A_1 at the jth stage. For example, if we compute the orbit of p_0 using the (not so random) sequence $(01\ 01\ 01 \ldots)$, we find that the x-coordinates of the points on the orbit are

---

*   For a proof, we refer the reader to M. Barnsley, *Fractals Everywhere*, Academic Press, 1988.

given by

$$x_1 = \frac{x_0}{3}$$

$$x_2 = \frac{x_0}{3^2} + \frac{2}{3}$$

$$x_3 = \frac{x_0}{3^3} + \frac{2}{3^2}$$

$$x_4 = \frac{x_0}{3^4} + \frac{2}{3^3} + \frac{2}{3}$$

$$x_5 = \frac{x_0}{3^5} + \frac{2}{3^4} + \frac{2}{3^2}$$

$$x_6 = \frac{x_0}{3^6} + \frac{2}{3^5} + \frac{2}{3^3} + \frac{2}{3}.$$

There is a pattern here. If n is even,

$$x_n = \frac{x_0}{3^n} + \frac{2}{3^{n-1}} + \frac{2}{3^{n-3}} + \frac{2}{3^{n-5}} + \cdots + \frac{2}{3},$$

while if n is odd,

$$x_n = \frac{x_0}{3^n} + \frac{2}{3^{n-1}} + \frac{2}{3^{n-3}} + \frac{2}{3^{n-5}} + \cdots + \frac{2}{9}.$$

We may amalgamate both of these sums as

$$x_n = \frac{x_0}{3^n} + \left(\frac{2s_1}{3^n} + \frac{2s_2}{3^{n-1}} + \frac{2s_3}{3^{n-2}} + \cdots + \frac{2s_n}{3} \right),$$

since $s_1 = s_3 = s_5 = \cdots = 0$ and $s_2 = s_4 = \cdots = 1$.

Note that, as $n \to \infty$, the first term in this sum tends to 0 while the remaining terms tend to an infinite series of the form

$$\sum_{i=1}^{\infty} \frac{t_i}{3^i}$$

where the t_i are alternately 0 and 2. The sequence x_n does not converge to a single number. Rather, the even terms of this sequence tend to one series and the odd to another. We have

$$\lim_{n \to \infty} x_{2n} = \lim_{n \to \infty} \left[\frac{x_0}{3^{2n}} + \left(\frac{2}{3^{2n-1}} + \frac{2}{3^{2n-3}} + \cdots + \frac{2}{3} \right) \right]$$

$$= \sum_{i=1}^{\infty} \frac{2}{3^{2i-1}} = \frac{3}{4}$$

and

$$\lim_{n \to \infty} x_{2n+1} = \lim_{n \to \infty} \left[\frac{x_0}{3^{2n+1}} + \left(\frac{2}{3^{2n}} + \frac{2}{3^{2n-2}} + \cdots + \frac{2}{3^2} \right) \right]$$

$$= \sum_{i=1}^{\infty} \frac{2}{3^{2i}} = \frac{1}{4}.$$

As we saw in Section 7.3, both of these points lie in the Cantor middle-thirds set.

In the general case, we may write

$$x_n = \frac{x_0}{3^n} + \left(\frac{2s_1}{3^n} + \frac{2s_2}{3^{n-1}} + \frac{2s_3}{3^{n-2}} + \cdots \right).$$

The first term in this sum again vanishes as $n \to \infty$, showing that, in the limit, this quantity is independent of x_0. The remaining terms in parentheses tend to an infinite series which is again of the form

$$\sum_{i=1}^{\infty} \frac{t_i}{3^i}$$

where t_i is either 0 or 2. As in Section 7.3 we recognize these series as giving numbers whose ternary expansion contains no digit equal to one. These, of course, correspond to points in the Cantor set.

Remark. In this example we could have played the chaos game on the x-axis instead of the plane. In this case the two linear contractions would be given by

$$A_0(x) = \frac{1}{3}x$$

$$A_1(x) = \frac{1}{3}x + \frac{2}{3}.$$

Note that the inverses of these two contractions are precisely the "microscopes" we used back in Section 14.2 to view the self-similarity of C.

We emphasize that the orbit does not in general converge to a single point in the Cantor set. Rather, as we randomly apply A_0 or A_1, the points on the orbit tend to move around the set, gradually visiting all regions in the set.

Example. Consider the five points

$$p_0 = \begin{pmatrix} 0 \\ 0 \end{pmatrix}, \ p_1 = \begin{pmatrix} 1 \\ 0 \end{pmatrix}, \ p_2 = \begin{pmatrix} 0 \\ 1 \end{pmatrix}, \ p_3 = \begin{pmatrix} 1 \\ 1 \end{pmatrix}, \ p_4 = \begin{pmatrix} 1/2 \\ 1/2 \end{pmatrix}.$$

Let A_i denote the linear contraction with the fixed point p_i and contraction factor $1/3$. The iterated function system generated by the A_i has an attractor which is the box fractal displayed in Figure 14.7. Note that, geometrically, each of the A_i contracts the unit square $0 \le x, y \le 1$ onto one of the five subsquares depicted in Figure 14.15.

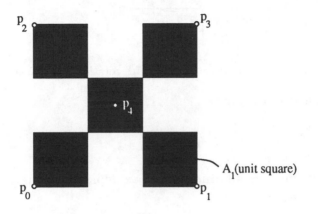

Fig. 14.15 Each A_i contracts the unit square s onto a subsquare.

There are a number of variations and generalizations of the concept of iterated function system. For example, we may assume that each contraction not only contracts points by a factor β toward the fixed point, but also rotates these points. To be specific, let θ be a nonzero angle. Then the function

$$A \begin{pmatrix} x \\ y \end{pmatrix} = \beta \cdot \begin{pmatrix} \cos \theta & -\sin \theta \\ \sin \theta & \cos \theta \end{pmatrix} \cdot \begin{pmatrix} x - x_0 \\ y - y_0 \end{pmatrix} + \begin{pmatrix} x_0 \\ y_0 \end{pmatrix}$$

leaves fixed the point

$$p = \begin{pmatrix} x_0 \\ y_0 \end{pmatrix}.$$

Any other point in the plane is first contracted by a factor of β toward p and then rotated by angle θ about p.

Example. Let $\beta = 0.9$ and $\theta = \pi/2$. Then

$$A \begin{pmatrix} x \\ y \end{pmatrix} = 0.9 \cdot \begin{pmatrix} 0 & -1 \\ 1 & 0 \end{pmatrix} \cdot \begin{pmatrix} x - 1 \\ y - 1 \end{pmatrix} + \begin{pmatrix} 1 \\ 1 \end{pmatrix}$$

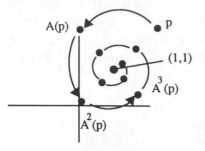

Fig. 14.16 An orbit of a contraction map with rotation $\theta = \pi/2$ about (1,1).

is a linear contraction that fixes

$$p_0 = \begin{pmatrix} 1 \\ 1 \end{pmatrix}$$

and rotates and contracts all other points as depicted in Figure 14.16.

Example. Let

$$p_0 = \begin{pmatrix} 0 \\ 0 \end{pmatrix} \quad p_1 = \begin{pmatrix} 1 \\ 0 \end{pmatrix} \quad p_2 = \begin{pmatrix} 0 \\ 1 \end{pmatrix}$$

and let A_i be the linear contraction with rotation $\pi/4$ and contraction factor $1/2$ about p_i. Then this iterated function system yields the attractor depicted in Figure 14.17.

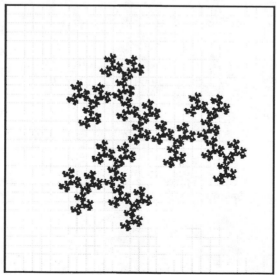

Fig. 14.17 An attractor generated by contraction and rotation.

There are many other possible generalizations of iterated function systems. For example, linear contractions need not contract all points equally. As discussed in linear algebra courses, there are more general types of linear transformations that contract the plane. Another type of iterated function system occurs if we apply the contractions with unequal probability. We refer to Barnsley's book* for a complete discussion of these topics.

Finally, it should be noted that the images generated by iterated function systems and other iterative processes may be quite lifelike. Indeed, one major branch of fractal geometry involves the use of iterated function systems that give accurate representations of natural objects such as clouds, ferns, mountain ranges, snowflakes, and the like.

14.8   Experiment: Iterated Function Systems

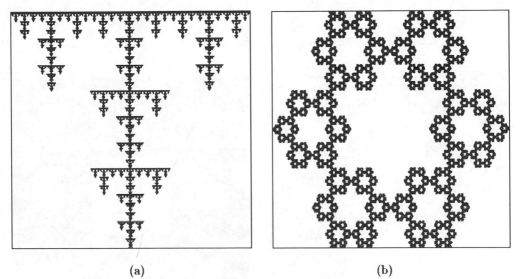

(a)   (b)

Fig. 14.18 Find the contractions that produce these attractors.

Goal: In this experiment you are asked to identify the iterated function systems that produced certain fractals as their attractors.

Procedure: First select a number of points in the plane as well as a contraction factor. Then, using the computer, randomly iterate the corresponding iterated function system and display its attractor. In Figures 14.18 and 14.19

---

*   *Fractals Everywhere*, op. cit.

we have displayed six different attractors. By creating different iterated function systems, your aim is to discover systems that generate the corresponding images.

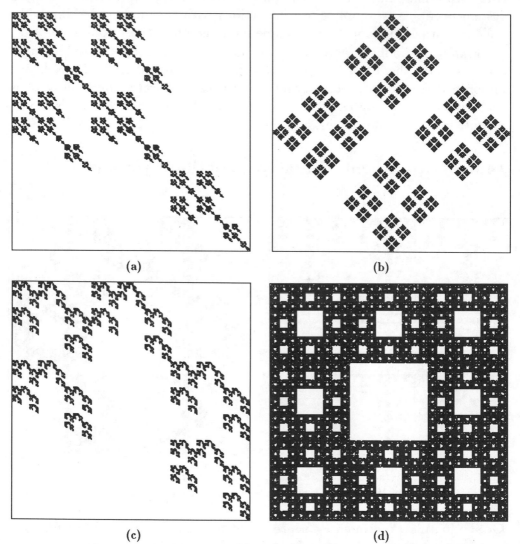

(a)

(b)

(c)

(d)

Fig. 14.19 Find the contractions that produce these attractors.

Results: In a brief essay, explain why you were led to choose the iterated function system that produced one of these images. Explain with pictures what each linear contraction in your system does to the unit square in the plane. Finally, what is the fractal dimension of each of these fractals?

Notes and Questions:

1. Each of the fractals in Figures 14.18 and 14.19 was obtained via iterated function systems without rotations. Figure 14.20 displays a fractal for which the corresponding iterated function system admits a nonzero rotation. How was this image produced?

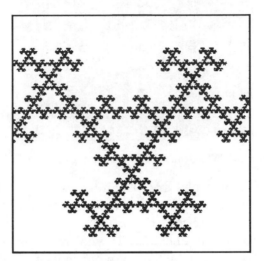

Figure 14.20 The contraction that produced this attractor involved nonzero rotations.

2. Note the darkened regions in Figure 14.21. What caused this? Identify the corresponding iterated function system.

Exercises

1. Without using the computer, predict the structure of the attractor generated by the iterated function system with contraction factor β and fixed points p_i:

    a. $\beta = 1/3$; $p_0 = \begin{pmatrix} 0 \\ 0 \end{pmatrix}$, $p_1 = \begin{pmatrix} 1 \\ 0 \end{pmatrix}$, $p_2 = \begin{pmatrix} 0 \\ 1 \end{pmatrix}$

    b. $\beta = 1/2$; $p_0 = \begin{pmatrix} 0 \\ 0 \end{pmatrix}$, $p_1 = \begin{pmatrix} 1 \\ 0 \end{pmatrix}$

    c. $\beta = 1/3$; $p_0 = \begin{pmatrix} 0 \\ 0 \end{pmatrix}$, $p_1 = \begin{pmatrix} 1 \\ 0 \end{pmatrix}$, $p_2 = \begin{pmatrix} 0 \\ 1 \end{pmatrix}$, $p_3 = \begin{pmatrix} 1 \\ 1 \end{pmatrix}$.

2. Give explicitly the iterated function system that generates the Cantor middle-fifths set. This set is obtained by the same process that generated the Cantor middle-thirds set, except that the middle fifth of each interval is removed at each stage. What is the fractal dimension of this set?

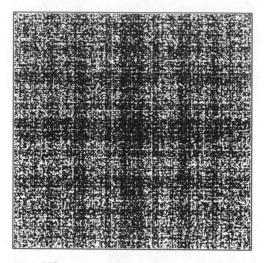

Figure 14.21 What caused the darker regions in this image?

3. Consider the set C obtained from the interval $[0, 1]$ by first removing the middle third of the interval and then removing the middle fifths of the two remaining intervals. Now iterate this process, first removing middle thirds, then removing middle fifths. The set C is what remains when this process is repeated infinitely. Is C a fractal? If so, what is its fractal dimension?

The following seven exercises deal with the Sierpinski right triangle (see Figure 14.5) generated by the following contractions:

$$A_0 \begin{pmatrix} x \\ y \end{pmatrix} = \frac{1}{2} \begin{pmatrix} x \\ y \end{pmatrix}$$

$$A_1 \begin{pmatrix} x \\ y \end{pmatrix} = \frac{1}{2} \begin{pmatrix} x - 1 \\ y \end{pmatrix} + \begin{pmatrix} 1 \\ 0 \end{pmatrix}$$

$$A_2 \begin{pmatrix} x \\ y \end{pmatrix} = \frac{1}{2} \begin{pmatrix} x \\ y - 1 \end{pmatrix} + \begin{pmatrix} 0 \\ 1 \end{pmatrix}$$

4. What are the fixed points for A_0, A_1, and A_2?

5. Show that $A_1^n \begin{pmatrix} x_0 \\ y_0 \end{pmatrix}$ converges to $\begin{pmatrix} 1 \\ 0 \end{pmatrix}$.

6. To which point does the sequence

$$A_2^2 \left(A_1^n \begin{pmatrix} x_0 \\ y_0 \end{pmatrix} \right)$$

converge?

7. Show that the sequence $A_1 \circ A_0^n \begin{pmatrix} x_0 \\ y_0 \end{pmatrix}$ converges to $\begin{pmatrix} 1/2 \\ 0 \end{pmatrix}$.

8. Show that the sequence

$$(A_1 \circ A_0)^n \begin{pmatrix} x_0 \\ y_0 \end{pmatrix}$$

accumulates on the two points

$$\begin{pmatrix} 1/3 \\ 0 \end{pmatrix} \quad \text{and} \quad \begin{pmatrix} 2/3 \\ 0 \end{pmatrix}.$$

9. On which points does the sequence

$$(A_2 \circ A_1)^n \begin{pmatrix} x_0 \\ y_0 \end{pmatrix}$$

accumulate?

10. Show that the sequence

$$(A_2 \circ A_1 \circ A_0)^n \begin{pmatrix} x_0 \\ y_0 \end{pmatrix}$$

accumulates on the points

$$\begin{pmatrix} 2/7 \\ 4/7 \end{pmatrix}, \begin{pmatrix} 1/7 \\ 2/7 \end{pmatrix}, \quad \text{and} \quad \begin{pmatrix} 4/7 \\ 1/7 \end{pmatrix}.$$

11. Consider the fractal generated by replacing a line segment with the smaller segments shown in Figure 14.22, where each new segment is exactly one-third as long as the original. Draw carefully the next two iterations of this process. What are the fractal and topological dimensions of the resulting fractal?

Fig. 14.22

12. Show that a set with topological dimension 0 is totally disconnected.

13. Can a fractal that is totally disconnected (topological dimension 0) have a fractal dimension larger than 1?

14. Show that the rational numbers form a subset of the real line that has topological dimension 0. What is the topological dimension of the set of irrationals?

15. Compute exactly the area of the Koch snowflake.

16. Show that the Koch curve may also be obtained by the sequence of removals shown in Figure 14.23.

Fig. 14.23 Construction of the Koch curve.

17. Consider the standard Pascal's triangle generated by binomial coefficients. In this triangle, replace each odd number by a black dot and each even number by a white dot. Describe the figure that results.

18. Rework exercise 17, this time replacing each number by a black dot if it is congruent to 1 mod 3 and a white dot otherwise. (That is, points that yield a remainder of 1 upon division by 3 are colored black.) Now describe the resulting figure. How does it compare with the figure generated in the previous exercise?

CHAPTER 15

Complex Functions

In this chapter we begin to study dynamical systems in the plane. We will consider here functions of a complex variable. This study, begun by the French mathematicians Gaston Julia and Pierre Fatou in the 1920's, received renewed impetus with the pioneering computer graphics work of Benoit Mandelbrot in 1980 and the subsequent mathematical investigations of Adrien Douady, John Hubbard, Dennis Sullivan, and others. In this chapter we will introduce the basic mathematical notions necessary to understand Julia sets for complex functions and the Mandelbrot set, our topics in the subsequent chapters.

15.1 Complex Arithmetic

We begin with a review of complex numbers. A *complex number* is a number of the form $x + iy$ where both x and y are real numbers and i is the imaginary number satisfying $i^2 = -1$. That is, i is, by definition, $\sqrt{-1}$. For example, $2 + 3i$ and $7 - 4i$ are both complex numbers. Similarly, both 7 and $7i$ are complex numbers, for we may write $7 = 7 + 0i$ and $7i = 0 + 7i$.

If $z = x + iy$ is a complex number, we call x the *real part* of z and y the *imaginary part*. The real number $\sqrt{x^2 + y^2}$ is called the *modulus* of z. We denote the modulus of z by $|z|$. We denote the set of all complex numbers by **C**.

The arithmetic operations of addition and multiplication are defined for complex numbers in the natural way. For example, we add two complex numbers by summing their real parts and their imaginary parts. Thus

$$(2 + 3i) + (4 + 5i) = 6 + 8i.$$

Multiplication is also defined in the natural manner, recalling that $i^2 = -1$. Thus

$$(x + iy) \cdot (u + iv) = xu + i^2 yv + ixv + iyu$$
$$= xu - yv + i(xv + yu).$$

For example,

$$(1 + 2i) \cdot (2 + 3i) = -4 + 7i$$
$$2i \cdot (1 + i) = -2 + 2i$$
$$4i \cdot 7i = -28.$$

Complex numbers may be depicted geometrically as points in the plane. We simply place the complex number $x + iy$ at the point (x, y) in the Cartesian plane. Thus i is placed at $(0,1)$ and $1 + i$ at $(1, 1)$, as shown in Figure 15.1.

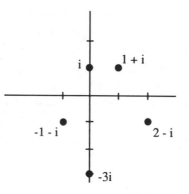

Fig. 15.1 Some complex numbers in the complex plane.

We call the plane with points identified as complex numbers the *complex plane*.

An alternative method of describing points in the complex plane is the *polar representation* of a complex number. Given a complex number $z = x + iy$, its polar representation is determined by the modulus of z,

$$|z| = \sqrt{x^2 + y^2},$$

which represents the distance from the origin to z, and the *polar angle* of z, which is the angle between the positive x-axis and the ray from 0 to z measured in the counterclockwise direction. The polar angle of z is sometimes called the *argument* of z. We often write $r = |z|$ and θ for the modulus and polar angle of z, the familiar polar coordinates of the point (x, y). Then the polar representation of $z = x + iy$ is $z = r \cos \theta + ir \sin \theta$, as shown in Figure 15.2. Recall Euler's Formula from elementary calculus:

$$e^{i\theta} = \cos \theta + i \sin \theta.$$

We thus see that any complex number may also be written in polar form as $z = re^{i\theta}$. For later use, we note that

$$|e^{i\theta}| = \sqrt{\cos^2(\theta) + \sin^2(\theta)} = 1.$$

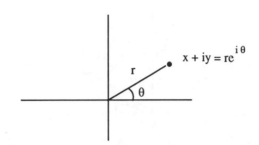

Fig. 15.2 The polar representation of $z = x + iy$ as $z = r \cos \theta + ir \sin \theta$.

Geometrically, addition of complex numbers is given by the parallelogram for addition of vectors in the plane. This is shown in Figure 15.3. In particular, this figure illustrates what will become a crucial fact for us later.

The Triangle Inequality. *If z and w are complex numbers, then*

1. $|z + w| \le |z| + |w|$, *and*
2. $|z - w| \ge |z| - |w|$.

The first of these inequalities follows immediately from Figure 15.3. The second is a consequence of the first, since

$$|z| = |z - w + w| \le |z - w| + |w|.$$

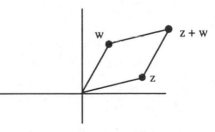

Fig. 15.3 Complex addition.

The polar representation of complex numbers makes it particularly easy to visualize complex multiplication. If $z = r(\cos\theta + i\sin\theta)$ and $w = \rho(\cos\phi + i\sin\phi)$, then the product zw is given by

$$z \cdot w = r\rho\left[(\cos\theta\cos\phi - \sin\theta\sin\phi) + i(\cos\theta\sin\phi + \cos\phi\sin\theta)\right]$$
$$= r\rho\left[\cos(\theta + \phi) + i\sin(\theta + \phi)\right].$$

Thus, to multiply two complex numbers, we simply multiply their moduli and add their polar angles as depicted in Figure 15.4.

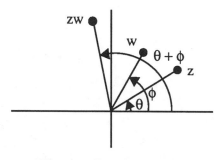

Fig. 15.4 Complex multiplication.

Division of complex numbers is defined in a roundabout way. First, given a complex number $z = x + iy$, we define its *complex conjugate* to be the complex number $\bar{z} = x - iy$. Note that $\bar{z}$ is simply the reflection of z through the x-axis. We have

$$z\bar{z} = (x + iy) \cdot (x - iy)$$
$$= x^2 + y^2$$
$$= |z|^2,$$

so $z\bar{z}$ is always a non-negative real number.

Given two complex numbers z and w, with $z \neq 0$, we therefore set

$$\frac{w}{z} = \frac{w \cdot \bar{z}}{z \cdot \bar{z}} = \frac{w \cdot \bar{z}}{|z|^2}.$$

Since the denominator of the right-hand side is real and nonzero, this definition of w/z makes sense. Note that this definition yields

$$\frac{w}{z} \cdot z = \frac{w \cdot \bar{z}}{|z|^2} \cdot z = \frac{w|z|^2}{|z|^2} = w$$

as expected.

15.2 Complex Square Roots

Given a complex number $z = x + iy$, there is no problem computing its square. Indeed,

$$z^2 = (x + iy) \cdot (x + iy)$$
$$= x^2 - y^2 + i(2xy).$$

In polar representation, if

$$z = r(\cos\theta + i\sin\theta),$$

then

$$z^2 = r^2(\cos 2\theta + i\sin 2\theta),$$

as we saw in the previous section. That is, squaring a complex number has the effect of squaring its modulus and doubling its polar angle.

In the sequel, we will often need to undo the squaring operation. That is, we need to know how to compute complex square roots. The polar representation of complex multiplication tells us how to do this. Given $z = r(\cos\theta + i\sin\theta)$ in polar form, the two square roots of z are

$$\pm\sqrt{r}(\cos(\theta/2) + i\sin(\theta/2)).$$

Note that each of these complex numbers gives z when squared. Geometrically, $\sqrt{z}$ is obtained by taking the square root of the modulus of z and halving the polar angle (the other square root is the negative of this number). Thus, for example

$$\sqrt{i} = \pm\left(\cos\frac{\pi}{4} + i\sin\frac{\pi}{4}\right) = \pm\left(\frac{1}{\sqrt{2}} + i\cdot\frac{1}{\sqrt{2}}\right),$$

since $|i| = 1$ and the polar angle of i is $\pi/2$.

Similarly,

$$\sqrt{1+i} = \pm 2^{1/4}\left(\cos\frac{\pi}{8} + i\sin\frac{\pi}{8}\right),$$

since $|1+i| = \sqrt{2}$ and the polar angle of $1+i$ is $\pi/4$.

The square root operation is depicted geometrically in Figure 15.5.

Later, we will need to compute the square roots of all points in a given region in the complex plane. Geometrically this is accomplished just as above. We simply halve the polar angle of all points in the region and take

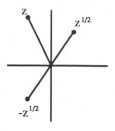

Fig. 15.5 Complex square roots.

the square root of their moduli. This yields one of the square roots of the region. The other square root is simply the negative of all of these points.

We specifically need to know how to compute the square root of all points that lie on a circle in the complex plane. There are three different possibilities, depending upon whether the origin lies inside, on, or outside this circle. To illustrate this, we begin with the circle of radius r centered at the origin. Clearly, the square root of this region is simply the circle of radius $\sqrt{r}$, also centered at the origin.

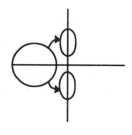

Fig. 15.6 Square root of a circle not containing 0.

To visualize this, it is helpful to imagine a particle travelling around the original circle, with another pair of particles simultaneously tracing the paths of the square roots. As the original particle makes one full loop around the circle, the square roots each traverse a semicircle, since we halve the corresponding polar angle.

If the origin lies outside of the circle, then the situation is quite different. The entire circle then lies within a wedge centered at the origin of the form $\theta_1 \leq \theta \leq \theta_2$, where $0 < |\theta_2 - \theta_1|, 2\pi$. As a particle traverses this circle, the corresponding square roots have polar angles that are constrained to lie within the wedge $\theta_1/2 \le \theta \le \theta_2/2$ and its reflection through the origin. As a consequence, the square root of this circle consists of two disjoint pieces, as shown in Figure 15.6.

The last case occurs when the origin actually lies on the circle. In this case, the square root is a curve that resembles a figure-eight (Fig. 15.7). The reason for this is that all points on the circle have exactly two square roots, except 0, which has just one. As a particle traverses the original circle, its square roots trace out each of the lobes of the figure-eight, meeting at the origin precisely when the original particle reaches 0.

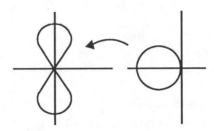

Fig. 15.7 Square root of a circle containing 0 is a figure-eight.

Figure 15.8 shows the square roots of some other regions in the plane.

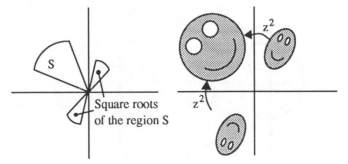

Fig. 15.8 Some regions in **C** and their square roots.

15.3. Linear Complex Functions

In this section we describe the dynamics of the simplest complex functions, namely, linear functions of the form $L_\alpha(z) = \alpha z$ where α is a complex number.

Note that L_α has a fixed point at $z = 0$. Our goal is to describe the orbits of nonzero points under iteration of L_α. Toward that end, we write $\alpha = \rho e^{i\psi}$. Let $z_0 = re^{i\theta}$. Then we have

$$z_1 = \rho e^{i\psi} \cdot re^{i\theta} = \rho re^{i(\psi+\theta)}$$
$$z_2 = \rho^2 e^{i \cdot 2\psi} \cdot re^{i\theta} = \rho^2 re^{i(2\psi+\theta)}$$
$$\vdots$$
$$z_n = \rho^n e^{i \cdot n\psi} re^{i\theta} = \rho^n re^{i(n\psi+\theta)}.$$

Thus, there are three possible cases. If $\rho < 1$, then $\rho^n \to 0$ as $n \to \infty$. Since

$$|e^{i(n\psi+\theta)}| = 1$$

for all n, it follows that $|z_n| \to 0$ as $n \to \infty$. Hence, if $\rho < 1$, all orbits of L_α tend to 0.

If $\rho > 1$, the opposite happens. Since $\rho^n \to \infty$ as $n \to \infty$, it follows that all nonzero orbits tend to infinity. In analogy with our analysis of real functions, we call 0 an *attracting fixed point* if $\rho < 1$ and a *repelling fixed point* if $\rho > 1$. Note that the polar angle $n\psi + \theta$ changes as n increases, provided $\psi \neq 0$. This means that orbits may spiral into or away from the origin, as shown in Figure 15.9.

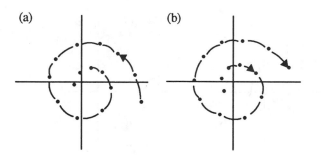

Fig. 15.9 Orbits of L_α where **(a)** $|\alpha| < 1$ and **(b)** $|\alpha| > 1$.

The neutral case (when $\rho = 1$) is more complicated than the analogous case for real linear maps. There are two important subcases depending upon the polar angle ψ. Let's write

$$\psi = 2\pi\tau.$$

The two different cases occur when τ is rational and τ is irrational.

Suppose first that $\tau = p/q$ where $p, q \in \mathbf{Z}$. Then we have

$$L_\alpha^q(re^{i\theta}) = re^{(2\pi ip/q)q+i\theta}$$
$$= re^{2\pi ip+i\theta}$$
$$= re^{i\theta}.$$

Hence each point $z_0 \neq 0$ is periodic with period q for L_α. Note that all points on the orbit of z_0 lie on the circle centered at 0 with radius $|z_0|$, as shown in Figure 15.10.

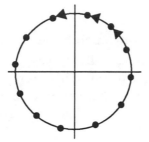

Fig. 15.10 Dynamics of L_α where $\alpha = e^{2\pi i/12}$.

When τ is irrational, the dynamics of L_α are quite different. There are no periodic points whatsoever for L_α (except 0), since, if $L_\alpha^k(z_0) = z_0$ for some k, we would have

$$re^{i\theta} = L_\alpha^k(re^{i\theta})$$
$$= re^{2\pi i\tau k+i\theta}.$$

Therefore, for some integer m, we must have

$$\theta + 2\pi m = \theta + 2\pi k\tau.$$

Hence $\tau = m/k$, which is a contradiction.

In fact, one can prove more. If τ is irrational and $\rho = 1$, the orbit of z_0 is a dense subset of the circle whose radius is $|z_0|$. This result is known as Jacobi's Theorem. To see why this is true, we must show that the orbit of z_0 enters any subarc of the circle of radius $|z_0|$ of length ϵ. To find such a point, choose an integer $k > 2\pi|z_0|/\epsilon$. The points $z_0, z_1, \ldots, z_k$ all lie on the circle of radius z_0 about 0 and they are all distinct. Since the circumference of this circle is $2\pi|z_0|$ and $2\pi|z_0|/k < \epsilon$, it follows that at least two points among the z_i are closer together than ϵ. Suppose the arc between z_j and z_ℓ has length less than ϵ and $j > \ell$.

Consider now the function $L_\alpha^{j-\ell}$. We have

$$L_\alpha^{j-\ell}(z_0) = e^{2\pi i\tau(j-\ell)}z_0,$$

so $L_\alpha^{j-\ell}$ simply rotates points by angle $2\pi\tau(j-\ell)$. Since

$$L_\alpha^{j-\ell}(z_\ell) = L_\alpha^{j-\ell}L_\alpha^\ell(z_0)$$
$$= L_\alpha^j(z_0)$$
$$= z_j,$$

it follows that this function rotates points on the circle a distance smaller than ϵ. Hence the points $L_\alpha^{j-\ell}(z_0), L_\alpha^{2(j-\ell)}(z_0), \ldots, L_\alpha^{n(j-\ell)}(z_0), \ldots$ are arranged around the circle with the distance between successive points no larger than ϵ. It follows that the orbit of z_0 must enter any subarc whose length is less than ϵ. Since ϵ was arbitrary, it follows that the orbit of z_0 is dense in the circle of radius $|z_0|$.

We may therefore summarize the dynamics of the linear complex maps as follows.

Proposition. *Suppose $L_\alpha(z) = \alpha z$ where $\alpha = \rho e^{2\pi i\tau}$.*

1. *If $\rho < 1$, all orbits tend to the attracting fixed point at 0.*
2. *If $\rho > 1$, all orbits tend to infinity, except that of 0, which is a repelling fixed point.*
3. *If $\rho = 1$:*
    a. *If τ is rational, all orbits are periodic.*
    b. *If τ is irrational, each orbit is dense on a circle centered at the origin.*

15.4   Calculus of Complex Functions

For a complex function $F(z)$, we define the complex derivative $F'(z)$ exactly as in the real case:

$$F'(z_0) = \lim_{z \to z_0} \frac{F(z) - F(z_0)}{z - z_0}.$$

While this definition looks straightforward, there really is a significant difference with real functions. In the real case, we need only check the limit $F'(x_0)$ in two directions, as x approaches x_0 from the left and from the right. In the complex case, we must check this limit in all possible directions in the plane, and these limits must all be the same.

Example. Let $F(z) = z^2 + c$. Then

$$F'(z_0) = \lim_{z \to z_0} \frac{(z^2 + c) - (z_0^2 - c)}{z - z_0}$$

$$= \lim_{z \to z_0} (z + z_0) \left(\frac{z - z_0}{z - z_0} \right)$$

$$= 2z_0.$$

Hence $F'(z_0)$ exists for each $z_0 \in \mathbf{C}$.

Example. Let $F(z) = \overline{z}$, the complex conjugate of z. At any point z_0, let's try to compute the limit $F'(z_0)$ along two straight lines through z_0. Let $\gamma(t) = z_0 + t$ and $\eta(t) = z_0 + it$. The line γ is parallel to the x-axis while η is parallel to the y-axis. We compute

$$\lim_{t \to 0} \frac{\overline{\gamma}(t) - \overline{z}_0}{\gamma(t) - z_0} = \lim_{t \to 0} \frac{\overline{z}_0 + t - \overline{z}_0}{z_0 + t - z_0} = 1$$

$$\lim_{t \to 0} \frac{\overline{\eta}(t) - \overline{z}_0}{\eta(t) - z_0} = \lim_{t \to 0} \frac{\overline{z}_0 - it - \overline{z}_0}{z_0 + it - z_0} = -1.$$

Since these limits are different, it follows that $F(z) = \overline{z}$ does not have a complex derivative.

Remark. The complex derivative is a much different object than the Jacobian matrix studied in multivariable calculus. For example, if we regard the function $F(z) = \overline{z}$ as the *real* function

$$F \begin{pmatrix} x \\ y \end{pmatrix} = \begin{pmatrix} x \\ -y \end{pmatrix},$$

then this function is differentiable in the real sense with Jacobian matrix

$$DF = \begin{pmatrix} 1 & 0 \\ 0 & -1 \end{pmatrix}.$$

The property of having a complex derivative is much stronger than having partial derivatives.

Example. Let $F(x + iy) = x + iy^2$. This function also does not possess a complex derivative. Again let $\gamma(t) = x_0 + t + iy_0$ and $\eta(t) = x_0 + i(y_0 + t)$. Along $\gamma(t)$ we have

$$\lim_{t \to 0} \frac{F(x_0 + t + iy_0) - F(x_0 + iy_0)}{t} = \lim_{t \to 0} \frac{t}{t} = 1,$$

whereas along $\eta(t)$ we have

$$\lim_{t \to 0} \frac{F(x_0 + i(y_0 + t)) - F(x_0 + iy_0)}{it}$$
$$= \lim_{t \to 0} \frac{i(y_0 + t)^2 - iy_0^2}{it}$$
$$= \lim_{t \to 0} \frac{2y_0 t + t^2}{t} = 2y_0.$$

Hence these limits are in general different and $F'(z_0)$ fails to exist.

The meaning of $F'(z_0)$ is best illustrated by linear functions of the form $F(z) = \alpha z$ where $\alpha \in \mathbf{C}$. As we saw in the previous section, F expands or contracts the plane by a factor of $|\alpha|$, and F rotates the plane by an amount equal to the polar angle of α. We easily compute $F'(z) = \alpha$ for all $z \in \mathbf{C}$. Hence we see that $|F'(z_0)|$ indicates the local expansion or contraction of F near z_0, while the polar angle of $F'(z_0)$ indicates the local rotation near z_0.

In particular this allows us to extend our notion of attracting and repelling fixed points to the complex plane.

Definition. Suppose F is a complex function with a fixed point z_0, that is, $F(z_0) = z_0$. Then
   1. The fixed point is *attracting* if $|F'(z_0)| < 1$.
   2. The fixed point is *repelling* if $|F'(z_0)| > 1$.
   3. The fixed point is *neutral* if $|F'(z_0)| = 1$.

Remarks: 1. Note that these definitions agree with our original definitions of these concepts for real functions in Chapter 5.
2. There is one significant difference between the complex case and the real case. For neutral fixed points, we have some new possibilities. Not

only may we have $F'(z_0) = \pm 1$, but we may also have $F'(z_0) = e^{i\theta}$. These types of neutral fixed points have nearby dynamics that may be extremely complicated. In fact, understanding the local dynamics near a neutral fixed point is the subject of intense recent research. Discussing these results here would take us well beyond the scope of this book. The interested reader may find some discussion of this topic in [Devaney, pp. 300-311].

3. Again as in the case of real functions, we extend the notions of attraction and repulsion to periodic points of period n by considering $|(F^n)'(z_0)|$ instead of $|F'(z_0)|$.

   As in the real case, there is a natural reason for the use of the terms "attracting" and "repelling" to describe these fixed points.

Attracting Fixed Point Theorem (Complex Case). *Suppose z_0 is an attracting fixed point for the complex function F. Then there is a disk D of the form $|z - z_0| < \delta$ about z_0 in which the following condition is satisfied: if $z \in D$, then $F^n(z) \in D$, and, moreover, $F^n(z) \to z_0$ as $n \to \infty$.*

Proof: Since $|F'(z_0)| < 1$, we may find $\delta > 0$ and $\mu < 1$ such that, if $|z - z_0| < \delta$, then

$$\left| \frac{F(z) - F(z_0)}{z - z_0} \right| = \frac{|F(z) - F(z_0)|}{|z - z_0|} < \mu < 1.$$

The disk D will be the disk of radius δ centered at z_0. Hence $|F(z) - z_0| = |F(z) - F(z_0)| < \mu|z - z_0|$ in D. This implies that $F(z)$ is closer to z_0 than z was by a factor of $\mu < 1$. Applying this result n times, we find

$$|F^n(z) - z_0| < \mu^n|z - z_0|.$$

This completes the proof.

   As in the real case we also have an analogous result for repelling fixed points.

Repelling Fixed Point Theorem (Complex Case). *Suppose z_0 is a repelling fixed point for the complex function F. Then there is a disk D of the form $|z - z_0| < \delta$ about z_0 in which the following condition is satisfied: if $z \in D$ but $z \neq z_0$, then there is an integer $n > 0$ such that $F^n(z) \notin D$.*

We will omit the proof of this fact, since it closely resembles the proof in the attracting fixed point case.

Example. Consider $Q_1(z) = z^2 + 1$. This function has a pair of fixed points given by

$$q_\pm = \frac{1 \pm i\sqrt{3}}{2}.$$

These fixed points are both repelling since

$$|Q_1'(q_\pm)| = |1 \pm i\sqrt{3}| = 4.$$

Compare the case of the real version of this function, which, as we know from Chapter 5, has no fixed or periodic points whatsoever.

For the remainder of this section, we consider only the complex quadratic functions $Q_c(z) = z^2 + c$ where $c \in \mathbf{C}$. Our goal is to show the following important fact.

Boundary Mapping Principle: *Suppose R is a closed set in the plane. If z_0 is an interior point* in R, then $Q_c(z_0)$ is an interior point of the image $Q_c(R)$.*

Remarks:
1. An equivalent way of formulating this principle is to say if z_1 belongs to the boundary of $Q_c(R)$, then the inverse images of z_1 lie in the boundary of R.
2. The Boundary Mapping Principle actually holds for any complex function that has a complex derivative everywhere, but we will not prove this here.
3. It is important to realize that real functions of the plane do not have this property. For example, consider $F(x + iy) = x + iy^2$. Under F, the plane is mapped onto the half-plane $\{(x, y) \mid y \geq 0\}$. So points in the interior of the plane are mapped to the boundary of the image, namely the x-axis. Recall that this function does not possess a complex derivative.

To see why the Boundary Mapping Principle holds, we introduce the notion of a "chunk" of a wedge about z_0. Suppose first that $z_0 \neq 0$ and that

---

*   By interior point we mean there is a small open disk about z_0 of the form $\{z \mid |z - z_0| < \delta\}$ that is completely contained in R.

$z_0 = r_0 e^{i\theta_0}$. Suppose $r_1 < r_0 < r_2$ and $\theta_1 < \theta_0 < \theta_2$. Then we define a *chunk* about z_0 to be a set of the form

$$W = \{re^{i\theta} \mid r_1 < r < r_2, \ \theta_1 < \theta < \theta_2\}$$

with $z_0 \in W$. Such a chunk is depicted in Figure 15.11. Note that z_0 lies in the interior of W.

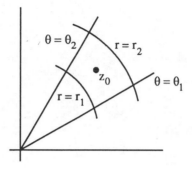

Fig. 15.11 A chunk of a wedge about z_0.

In the special case where $z_0 = 0$, we take the chunk to be a small disk centered at 0.

For the squaring function $Q_0(z) = z^2$, we note that chunks are always mapped to chunks. Indeed, the image under Q_0 of the chunk W above is given by

$$Q_0(W) = \{re^{i\theta} \mid r_1^2 < r < r_2^2, 2\theta_1 < \theta < 2\theta_2\}.$$

Note that $Q_0(W)$ contains $Q_0(z_0)$ in its interior.

In the case $c \neq 0$, Q_c takes chunks to regions that are chunks which have been translated by the complex number c (see Fig. 15.12).

This observation allows us to see why the Boundary Mapping Principle holds. Given z_0 inside a region R, we simply choose a small enough chunk about z_0 that lies entirely inside R. The image of this chunk then lies inside $Q_c(R)$. By the above observation, $Q_c(z_0)$ lies in the image chunk's interior, and therefore $Q_c(z_0)$ lies in the interior of $Q_c(R)$.

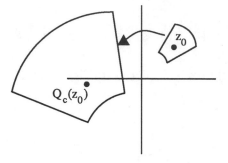

Fig. 15.12 Q_c takes a chunk about z_0 to one that has been translated to $Q_c(z_0)$.

Exercises

1. Compute the following:
   a. $(3 + 7i)^2$
   b. $(4 - 2i)^3$
   c. $(7 + 2i)(5 - 3i)(4i)$
   d. $\dfrac{3 + 2i}{6 - 5i}$
   e. $\dfrac{1}{(3 + 2i)^2}$

2. Find the polar representation of each of the following complex numbers:
   a. $-7i$
   b. -6
   c. $2 + 2i$
   d. $-2 + 2i$
   e. $1 + \sqrt{3}i$
   f. $-1 + \sqrt{3}i$

3. Find the complex square roots of each of the complex numbers in the previous exercise.

4. What is the formula for the quotient of two complex numbers given in polar representation?

5. Let $L_\alpha(z) = \alpha z$. Sketch the orbit of 1 in the plane for each of the following values of α:
   a. $\alpha = i/2$

 b. $\alpha = 2i$

 c. $\alpha = 1 + \sqrt{3}i$

 d. $\alpha = i$

 e. $\alpha = e^{2\pi i/9}$

 f. $\alpha = e^{\sqrt{2}\pi i}$

6. Prove that the complex function $F(z) = \alpha z + \beta$ where α and β are complex is conjugate to a linear function of the form $L_\gamma(z) = \gamma z$. Determine γ in terms of α and β. What happens when $\alpha = 1$?

7. For which of the following complex functions does the complex derivative exist?

 a. $F(x + iy) = (x + iy)^3$

 b. $F(x + iy) = x^2 + iy^2$

 c. $F(z) = |z|$

 d. $F(z) = \overline{z}^2 + c$

 e. $F(x + iy) = ix - y$

 f. $F(z) = 2z(i - z)$

 g. $F(z) = z^3 + (i + 1)z$

8. Find all fixed points for each of the following complex functions and determine whether they are attracting, repelling, or neutral.

 a. $Q_2(z) = z^2 + 2$

 b. $F(z) = z^2 + z + 1$

 c. $F(z) = iz^2$

 d. $F(z) = -1/z$

 e. $F(z) = 2z(i - z)$

 f. $F(z) = -iz(1 - z)/2$

 g. $F(z) = z^3 + (i + 1)z$

9. Show that $z_0 = -1 + i$ lies on a cycle of period 2 for $Q_i(z) = z^2 + i$. Is this cycle attracting, repelling, or neutral?

10. Show that $z_0 = e^{2\pi i/3}$ lies on a cycle of period 2 for $Q_0(z) = z^2$. Is this cycle attracting, repelling, or neutral?

11. Show that $z_0 = e^{2\pi i/7}$ lies on a cycle of period 3 for $Q_0(z) = z^2$. Is this cycle attracting, repelling, or neutral?

12. Show that the Boundary Mapping Principle holds for $F_c(z) = z^d + c$.

13. Does the Boundary Mapping Principle hold for $F(z) = \overline{z}$? Why or why not?

14. Does the Boundary Mapping Principle hold for $F(x + iy) = x^2 + iy^2$? Why or why not?

15. Give an example of a region R in the plane that has the property that, under $Q_0(z) = z^2$, there is a boundary point of R that is mapped into the interior of $Q_0(R)$. Does this contradict the Boundary Mapping Principle? Why or why not?

16. Prove the Repelling Fixed Point Theorem in the complex case.

CHAPTER 16

The Julia Set

The aim of this chapter is to introduce the notion of the Julia set of a complex function. As we will see, the Julia set is the place where all of the chaotic behavior of a complex function occurs. Keeping with our earlier philosophy, we will consider here only quadratic functions of the form

$$Q_c(z) = z^2 + c.$$

Here both z and c are complex numbers. In Chapter 18 we will consider a few other complex functions. We begin by revisiting three old friends.

16.1   The Squaring Function

The dynamics of the squaring function $Q_0(z) = z^2$ in the complex plane is especially easy to understand. Let's write $z_0 = re^{i\theta}$. Then the orbit of z_0 under Q_0 is given by

$$z_0 = re^{i\theta}$$
$$z_1 = r^2 e^{i(2\theta)}$$
$$z_2 = r^4 e^{i(4\theta)}$$
$$\vdots$$
$$z_n = r^{2^n} e^{i(2^n\theta)}$$
$$\vdots$$

This means that there are three possible fates for the orbit of z_0. If $r < 1$, we have

$$r^{2^n} \to 0 \text{ as } n \to \infty.$$

Hence $|Q_0^n(z_0)| \to 0$ as $n \to \infty$. As in the real case, $Q_0(0) = 0$ and $Q_0'(0) = 0$, so 0 is an attracting fixed point. On the other hand, if $r > 1$, then we have

$$r^{2^n} \to \infty \text{ as } n \to \infty,$$

so $|Q_0^n(z_0)| \to \infty$ as $n \to \infty$ in this case.

The intermediate case is $r = 1$, the unit circle. If $|z_0| = 1$, then $|Q_0(z_0)| = 1$ as well. Thus Q_0 preserves the unit circle in the sense that the image of any point on the unit circle also lies on the unit circle.

On the unit circle, the squaring map is the same as our old friend the doubling map. For if $z_0 = e^{i\theta}$ lies on the circle, we have $Q_0(z_0) = e^{i(2\theta)}$. That is, if we specify a point on the circle by giving its argument θ, then the image of this point has argument 2θ. So Q_0 simply doubles angles on the unit circle. As earlier, this means that Q_0 is chaotic on the unit circle. Our earlier proof of this works in this case, but let's prove a portion of this again using planar arguments.

To see that periodic points are dense, we must produce a periodic point in any arc of the form $\theta_1 < \theta < \theta_2$. That means we must find an n and θ so that

$$Q_0^n(e^{i\theta}) = e^{i\theta}$$

with $\theta_1 < \theta < \theta_2$. But

$$Q_0^n(e^{i\theta}) = e^{i \cdot 2^n \theta}$$

so we must solve

$$e^{i(2^n \theta)} = e^{i\theta}.$$

The angle θ solves this equation provided

$$2^n \theta = \theta + 2k\pi$$

for some integers k, n. That is, θ must satisfy

$$\theta = \frac{2k\pi}{2^n - 1}$$

where both k, n are integers. If we fix n and let k be an integer with $0 \leq k < 2^n - 1$, then the complex numbers with arguments $2k\pi/(2^n - 1)$ are evenly distributed around the circle as shown in Figure 16.1, with arclength

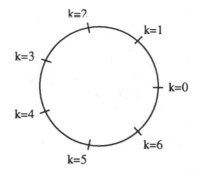

Fig. 16.1 Points on the unit circle with arguments $\theta = 2k\pi/(2^3 - 1)$.

$2\pi/(2^n - 1)$ between successive points. This follows since these points are just the $(2^n - 1)$st roots of unity on the unit circle.

If we now choose n so that

$$\frac{2\pi}{2^n - 1} < \theta_2 - \theta_1,$$

we guarantee that there is at least one point with argument $2k\pi/(2^n - 1)$ between θ_1 and θ_2. This point is therefore periodic with period n.

Transitivity follows as before. Any open arc $\theta_1 < \theta < \theta_2$ on the circle is doubled in arclength by Q_0. Hence Q_0^n magnifies arclengths by a factor of 2^n. This means that we may choose n large enough that the image of the arc $\theta_1 < \theta < \theta_2$ under Q_0^n covers the entire circle and therefore any other arc. This proves transitivity as well as sensitive dependence, for it shows that we may find nearby points that are eventually mapped to diametrically opposed points on the circle. To summarize, we have given a complete orbit analysis of Q_0 on the complex plane.

Theorem. *The squaring map $Q_0(z) = z^2$ is chaotic on the unit circle. If $|z| < 1$ then $|Q_0^n(z)| \to 0$ as $n \to \infty$. If $|z| > 1$ then $|Q_0^n(z)| \to \infty$ as $n \to \infty$.*

The squaring map is extremely sensitive to initial conditions in the following sense. Let z_0 be a point on the unit circle. Given any open ball about z_0, we may always find inside this ball a small chunk of a wedge of the form

$$W = \{re^{i\theta} \mid r_1 < r < r_2, \ \theta_1 < \theta < \theta_2\}$$

with $r_1 < 1 < r_2$. Such a chunk is depicted in Figure 16.2.

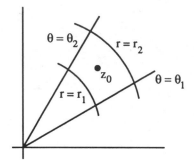

Fig. 16.2 A chunk of a wedge about z_0.

Now the image of W is a new chunk given by

$$Q_0(W) = \{re^{i\theta} \mid r_1^2 < r < r_2^2, \ 2\theta_1 < \theta < 2\theta_2\}.$$

Since $r_1 < 1 < r_2$, we have $r_1^2 < r_1 < 1 < r_2 < r_2^2$. Hence $Q_0(W)$ is a chunk whose inner and outer radii are larger than those of W, and whose total polar angle is double that of W. Continuing, we see that each application of Q_0 has the same effect so that the size of $Q_0^n(W)$ grows with n. Eventually, the polar angle of $Q_0^n(W)$ exceeds 2π so that, for n sufficiently large, $Q_0^n(W)$ is the annular region determined by

$$r_1^{2^n} < r < r_2^{2^n}.$$

Further applications of Q_0 show that the inner radius of $Q_0^n(W)$ tends to zero while the outer radius tends to infinity. We therefore see that

$$\bigcup_{n=0}^{\infty} Q_0^n(W) = \mathbf{C} - \{0\}.$$

That is, the orbits of points in W eventually reach any point in $\mathbf{C}$, except 0 (see Figure 16.3). This is extreme sensitive dependence: arbitrarily close to any point on the unit circle there is a point whose orbit includes any other point in the plane, with one exception, the origin.

We close this section with several definitions.

Definition. The orbit of z under Q_c is *bounded* if there exists K such that $|Q_c^n(z)| < K$ for all n. Otherwise, the orbit is *unbounded*.

For the squaring function, the orbit of any point inside and on the unit circle is bounded; points outside the unit circle have unbounded orbits.

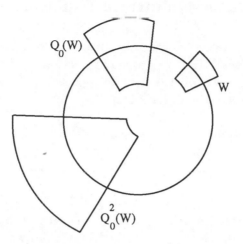

Fig. 16.3 Q_0 magnifies the chunk W

Definition. The orbit of z under Q_c is *supersensitive* if any open ball B about z has the property that

$$\bigcup_{n=0}^{\infty} Q_c^n(B)$$

is all of **C**, with the exception of at most one point.

For the squaring function, only points on the unit circle have supersensitive orbits. For if $|z| < 1$, we may choose a small open ball about z that lies completely inside the unit circle. Then the orbit of any point in this ball never leaves the set $\{z \mid |z| < 1\}$. If $|z| > 1$, we may similarly find a small open ball outside $\{z \mid |z| \le 1\}$ whose orbit remains outside of this region.

Definition. The *filled Julia set* of Q_c is the set of points whose orbits are bounded. The *Julia set* of Q_c is the boundary of the filled Julia set.

We denote the filled Julia set by K_c and the Julia set by J_c. For the squaring map we have thus shown that

$$K_0 = \{z \mid |z| \le 1\}$$
$$J_0 = \{z \mid |z| = 1\}.$$

Note also that it is precisely the points in J_0 whose orbits are supersensitive and that Q_0 is chaotic on J_0.

16.2   The Chaotic Quadratic Function

Now we turn to a second example from the complex quadratic family, $Q_{-2}(z) = z^2 - 2$. Recall that we proved in Section 10.2 that the real version of this function is chaotic on the interval $-2 \leq x \leq 2$. As we will see here, this interval is precisely the Julia set of Q_{-2}.

Theorem. *The quadratic function $Q_{-2}(z) = z^2 - 2$ on $\mathbf{C} - [-2, 2]$ is conjugate to the squaring function $Q_0(z) = z^2$ on $\{z \,|\, |z| > 1\}$. Consequently, if z does not lie in the closed interval $[-2, 2]$, then the orbit of z under Q_{-2} tends to infinity.*

Proof: Consider the function $H(z) = z + \frac{1}{z}$ defined on the region $R = \{z \,|\, |z| > 1\}$. H is one-to-one on R, for if $H(z) = H(w)$, then we have

$$z + \frac{1}{z} = w + \frac{1}{w},$$

so that, after some algebra,

$$zw = 1.$$

If $|z| > 1$ then it follows that $|w| = \frac{1}{|z|} < 1$ so $w \notin R$. Therefore, there is at most one point in R that is mapped by H to a point in $\mathbf{C}$. Also, H maps R onto $\mathbf{C} - [-2, 2]$. To see this, we choose $w \in \mathbf{C}$. We solve $H(z) = w$ via the quadratic formula, finding two solutions,

$$z_{\pm} = \frac{1}{2} \left(w \pm \sqrt{w^2 - 4} \right).$$

Since $z_{+} z_{-} = 1$, it follows that one of z_{+} or z_{-} lies in R or else both lie on the unit circle. In this latter case, however, it is easy to check that $H(z_{+}) = H(z_{-}) \in [-2, 2]$. This shows that H takes R in one-to-one fashion onto $\mathbf{C} - [-2, 2]$.

Finally, an easy computation shows that

$$H(Q_0(z)) = Q_{-2}(H(z))$$

for all z. Hence Q_{-2} on $\mathbf{C} - [-2, 2]$ is conjugate to Q_0 on R. Since all orbits of Q_0 tend to infinity in R, it follows that all orbits of Q_{-2} also tend to infinity in $\mathbf{C} - [-2, 2]$. This completes the proof.

Corollary. $J_{-2} = K_{-2} = [-2, 2]$.

Using the conjugacy H, we see also that the orbit of any point in J_{-2} is supersensitive. Indeed, if we consider the inverse image of any open ball about a point in J_{-2} under H, then we know that the squaring function Q_0 eventually smears this set over the entire plane. Hence, via the conjugacy, the same is true for Q_{-2}.

Remarks:

1. The conjugacy H looks different but is actually the same as the semi-conjugacy we encountered in Section 10.2. Indeed, if $z = e^{i\theta}$, then

$$H(z) = e^{i\theta} + e^{-i\theta}$$
$$= (\cos\theta + i\sin\theta) + (\cos\theta - i\sin\theta)$$
$$= 2\cos\theta,$$

which gave us the semi-conjugacy between z^2 and $z^2 - 2$ on the Julia set of each.

2. It is an interesting exercise to check that H maps straight rays perpendicular to the unit circle to (possibly degenerate) hyperbolas that have vertices in the interval $[-2, 2]$ as depicted in Figure 16.4.

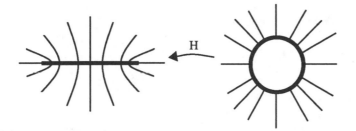

Fig. 16.4 H maps straight rays to hyperbolas.

3. The filled Julia set of Q_{-2} is the interval $[-2, 2]$ on the real line. The boundary of this set *in the plane* is itself; that is why $J_{-2} = K_{-2}$.

16.3   Cantor Sets Again

In this section we will consider the quadratic functions $Q_c(z) = z^2 + c$ in case $|c| > 2$. Recall that, for real quadratic functions $Q_c(x) = x^2 + c$, if $c < -2$, then there is an invariant Cantor set on which Q_c is chaotic. On the other hand, when $c > 2$, all orbits of Q_c on the real line simply tend to infinity. We will see, however, that these two subfamilies are really quite similar when viewed as dynamical systems in the complex plane. In fact, using techniques analogous to those introduced when we discussed symbolic dynamics in Chapter 9, it can be shown that:

Theorem. *Suppose* $|c| > 2$. *Then* $J_c = K_c$ *is a Cantor set. Moreover, the quadratic map* Q_c, *when restricted to* J_c, *is conjugate to the shift map on two symbols.*

We will not provide all of the details of the proof here, since some of the ideas involve advanced topics in complex analysis and others duplicate arguments we gave in detail in Chapter 9. We will, however, show how the Cantor set construction arises in this more general setting of the complex plane.

Our first observation is that the filled Julia set of Q_c when $|c| > 2$ is contained entirely inside the disk $|z| < |c|$.

Theorem (The Escape Criterion). *Suppose* $|z| \geq |c| > 2$. *Then we have* $|Q_c^n(z)| \to \infty$ *as* $n \to \infty$.

Proof: By the Triangle Inequality, we have

$$|Q_c(z)| \geq |z|^2 - |c|$$
$$\geq |z|^2 - |z| \quad \text{since } |z| \geq |c|$$
$$= |z|(|z| - 1).$$

Since $|z| > 2$, there is $\lambda > 0$ such that $|z| - 1 > 1 + \lambda$. Consequently

$$|Q_c(z)| > (1 + \lambda)|z|.$$

In particular, $|Q_c(z)| > |z|$, so we may apply the same argument repeatedly to find

$$|Q_c^n(z)| > (1 + \lambda)^n|z|.$$

Thus the orbit of z tends to infinity. This completes the proof.

This is an important theorem. Its corollaries will play an important role in the sequel. We first note that, if $|c| > 2$, then $|Q_c(0)| = |c| > 2$. Hence the orbit of 0, the critical point, necessarily escapes to infinity if $|c| > 2$.

Corollary 1. *Suppose* $|c| > 2$. *Then the orbit of* 0 *escapes to infinity under* Q_c.

The proof of the escape criterion actually gives us a little more information. In the proof, we only used the facts that $|z| \geq |c|$ and $|z| > 2$. Hence we have the following refinement of the Escape Criterion:

Corollary 2. *Suppose* $|z| > \max\{|c|, 2\}$. *Then* $|Q_c^n(z)| > (1 + \lambda)^n |z|$ *and so* $|Q_c^n(z)| \to \infty$ *as* $n \to \infty$.

As an additional note, we observe that if $|Q_c^k(z)| > \max\{|c|, 2\}$ for some $k \geq 0$, then we may apply this corollary to $Q_c^k(z)$ to find:

Corollary 3. *Suppose for some* $k \geq 0$ *we have* $|Q_c^k(z)| > \max\{|c|, 2\}$. *Then* $|Q_c^{k+1}(z)| > (1 + \lambda)|Q_c^k(z)|$, *so* $|Q_c^n(z)| \to \infty$ *as* $n \to \infty$.

Note that this corollary gives us an algorithm for computing the filled Julia set of Q_c for any c. Given any point z satisfying $|z| \leq |c|$, we compute the orbit of z. If, for some n, $Q_c^n(z)$ lies outside the circle of radius $\max\{|c|, 2\}$, we are guaranteed that the orbit escapes. Hence z is not in the filled Julia set. On the other hand, if $|Q_c^n(z)|$ never exceeds this bound, then z is by definition in K_c. We will make extensive use of this algorithm* in the next section.

For the remainder of this section, we restrict attention to the case $|c| > 2$. Let D denote the closed disk $\{z \mid |z| \leq |c|\}$. The filled Julia set of Q_c is given by

$$\bigcap_{n \geq 0} Q_c^{-n}(D)$$

where $Q_c^{-n}(D)$ is the preimage of D under Q_c^n, that is,

$$Q_c^{-n}(D) = \{z \mid Q_c^n(z) \in D\}.$$

This follows since, if $z \notin \cap_{n \geq 0} Q_c^{-n}(D)$, then there exists $k \geq 0$ such that $Q_c^k(z) \notin D$ and so the orbit of z tends to infinity by Corollary 3. Thus we need only understand the infinite intersection $\cap_{n \geq 0} Q_c^{-n}(D)$ to understand K_c.

---

*   It is perhaps better to call this a *semi-algorithm*, since we cannot determine whether a point actually lies in K_c in finite time.

Toward that end, let C denote the circle of radius $|c|$ centered at the origin. Note that C is the boundary of D. The first question is: what does $Q_c^{-1}(C)$ look like? We obtain this set by first subtracting c from any point in C. This has the effect of translating the circle so that it is centered at $-c$ and in fact passes through the origin. We then compute the complex square root of all points on this new circle. As we explained in Chapter 15, this yields a figure-eight curve as shown in Figure 16.5. Note that $Q_c^{-1}(C)$ is completely contained in the interior of D since we already know that any point on or outside of C is mapped farther away from the origin by Q_c. Moreover, by the Boundary Mapping Principle, $Q_c^{-1}(D)$ is precisely this figure-eight together with its two interior "lobes." Let us denote these lobes by I_0 and I_1 as shown in Figure 16.5. Note that I_0 and I_1 are symmetrically located about the origin and that Q_c maps each of them onto D in one-to-one fashion.

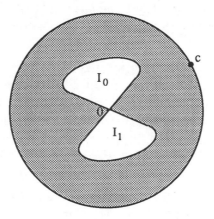

Fig. 16.5 Construction of I_0 and I_1.

Since Q_c is one-to-one on both I_0 and I_1, and since $Q_c^{-1}(C)$ is contained in the interior of D, it follows that $Q_c^{-2}(C)$ is a pair of smaller figure-eights, one contained in the interior of I_0 and one in the interior of I_1. Again invoking the Boundary Mapping Principle, we find that $Q_c^{-2}(D)$ consists of these two figure-eights together with their four lobes. We define

$$I_{00} = \{z \in I_0 \,|\, Q_c(z) \in I_0\}$$
$$I_{01} = \{z \in I_0 \,|\, Q_c(z) \in I_1\}$$
$$I_{10} = \{z \in I_1 \,|\, Q_c(z) \in I_0\}$$
$$I_{11} = \{z \in I_1 \,|\, Q_c(z) \in I_1\}.$$

Thus, $Q_c^{-2}(D)$ consists of a figure-eight in I_0 together with its lobes I_{00} and I_{01} and another figure-eight in I_1 with lobes I_{10} and I_{11} (see Figure 16.6). Note the similarity of this construction with the symbolic dynamics construction in Chapter 9.

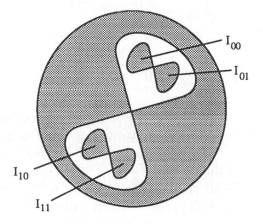

Fig. 16.6 Construction of I_{00}, I_{01}, I_{10}, and I_{11}.

Continuing, we see that $Q_c^{-n}(D)$ consists of 2^{n-1} figure-eights together with 2^n lobes bounded by these curves. Let $s_0 s_1 \ldots s_n$ be a string of 0's or 1's. We define

$$I_{s_0 s_1 \ldots s_n} = \{ z \in D \mid z \in I_{s_0}, Q_c(z) \in I_{s_1}, \ldots, Q_c^n(z) \in I_{s_n} \}$$

and note that this is precisely the definition we used to discuss the real dynamics of Q_c when $c < -2$ in Chapter 9.

The arguments used there in the proof of the Conjugacy Theorem in Section 9.4 may be mimicked here to show that $I_{s_0 s_1 \ldots s_n}$ is a closed lobe that is contained in the interior of the lobe $I_{s_0 s_1 \ldots s_{n-1}}$. Hence the $I_{s_0 s_1 \ldots s_n}$ form a nested intersection of closed sets. Therefore

$$\bigcap_{n \geq 0} I_{s_0 s_1 \ldots s_n}$$

is a nonempty set. This fact is analogous to the Nested Intersection Property discussed in Appendix A.3. Thus, if $z \in \cap_{n \geq 0} I_{s_0 s_1 \ldots s_n}$, then $Q_c^k(z) \in D$ for all k. Thus $z \in K_c$.

An infinite intersection of figure-eights and their lobes given by

$$\bigcap_{n \geq 0} I_{s_0 s_1 \ldots s_n}$$

is called a *component* of K_c. Note that any two components of K_c are necessarily disjoint.

Conversely, any $z \in K_c$ must lie in one of these components. Consequently, we may associate an infinite string of 0's and 1's to any such z via the rule

$$S(z) = s_0 s_1 s_2 \ldots$$

provided

$$z \in \bigcap_{n \geq 0} I_{s_0 s_1 \ldots s_n}.$$

This string of 0's and 1's is identical to the itinerary defined in Chapter 9. We have therefore shown that there is a natural correspondence between points in the sequence space on two symbols and the components of K_c. Similar arguments to those in Chapter 9 show that this correspondence is in fact continuous. A more difficult result is the fact that each of these components is indeed a single point. Recall that, even in the real case, we needed to make additional assumptions (on the size of the derivative) to prove this. Using more sophisticated techniques from complex analysis, the same result can be proved in this case. We will not, however, go into the details here. The basic idea is clear: the filled Julia set is a nested intersection of figure-eight curves and their lobes as shown in Figure 16.7.

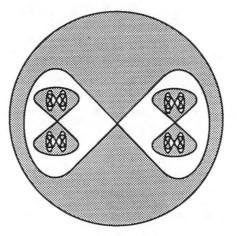

Fig. 16.7 The Julia set is a nested intersection of figure-eights when $|c| > 2$.

Remarks:

1. Accepting the fact that each component of K_c is a point when $|c| > 2$, we see that the Julia set and the filled Julia set are identical.

2. Figure 16.7 indicates that Q_c is also extremely sensitive to initial conditions on its Julia set when $|c| > 2$. Indeed, given any $z \in K_c$ and a small ball containing z in its interior, we may always choose k so large that the lobe $I_{s_0 s_1 \ldots s_k}$ containing z is also contained in this ball. But then Q_c^k maps this lobe onto the entire disk D, and subsequent iterations expand this disk further. Eventually, any point in the plane lies in the image of this lobe under a sufficiently high iterate of Q_c. Hence Q_c is supersensitive on J_c.

3. We emphasize the fact that, when $|c| > 2$, the orbit of 0 tends to infinity. This will become important in the next chapter when we describe the Mandelbrot set.

16.4   Computing the Filled Julia Set

In the previous sections we discussed three very special examples of filled Julia sets. As we will see, the typical complex quadratic function has a filled Julia set that is much more interesting than those described above. In this section we will again turn to the computer to determine experimentally the shapes of various filled Julia sets.

The easiest way to compute the filled Julia set is to use the definition of K_c. We consider a rectangular grid of points in some region in the plane. For each point in this grid, we compute the corresponding orbit and ask whether or not this orbit tends to infinity. If the orbit does not escape, then our original point is in K_c, so in the pictures in this section we will color the original point black. If the orbit escapes, then we leave the original point white. The only question is how we determine whether the orbit escapes to infinity. For this we can use the escape criterion in the previous section.

Algorithm for the Filled Julia Set: *Choose a maximum number of iterations, N. For each point z in the grid, compute the first N points on the orbit of z. If $|Q_c^i(z)| > \max\{|c|, 2\}$ for some $i \leq N$, then stop iterating and color z white. If $|Q_c^i(z)| \leq \max\{|c|, 2\}$ for all $i \leq N$, then color z black. White points have orbits that escape, whereas black points do not, at least for the first N iterations. So the black points yield an approximation to the filled Julia set.*

Remarks:

1. This algorithm is not foolproof. There may be points that take a larger number of iterations to exceed the bound $\max\{|c|, 2\}$ than our maximum number of iterations, N. In this case these points will be colored black although they are not in K_c.

2. Despite this fact, it is usually best to keep the maximum number of iterations low when using this algorithm. Thirty to sixty iterations are usually enough to give a very good approximation of K_c, except in cases where there is an indifferent periodic point in K_c. For magnifications of portions of the filled Julia set, you will need to increase the number N.

3. For the color plates of filled Julia sets we use a slightly different algorithm. The filled Julia set is still colored black. Points whose orbits escape are assigned colors depending upon the number of iterations necessary for the orbit to exceed $\max\{|c|, 2\}$. Red points have orbits that escape most quickly. Violet points have orbits that escape slowly. In between, points are colored shades of orange, yellow, green, blue, and indigo in increasing order of the number of iterations necessary to escape.

We have included a program that carries out this algorithm in Appendix B. We encourage you to use this program to experiment with the filled Julia sets of Q_c for a number of c-values. You will see that these sets assume a great variety of interesting shapes. For reasons that will become clear in the next chapter, we suggest that you choose only c-values that satisfy $|c| \leq 2$. We know already that, when $|c| > 2$, K_c is a Cantor set.

For the remainder of this section, we simply summarize some of the observations that you will make using the experiments in this chapter.

Observation 1: *For different c-values, K_c assumes a wide variety of shapes. Often, K_c consists of a large connected set in the plane. Some of these regions are displayed in Figure 16.8.*

A natural question is how to classify or understand all of these interesting shapes. We will see in the next section that it is the Mandelbrot set that provides a "dictionary" of all of these Julia sets.

The filled Julia sets in Figure 16.8 are remarkably different from those we encountered earlier in this chapter. Unlike the filled Julia sets for $c = 0$ (a disk) and $c = -2$ (an interval), for other values of c, K_c appears to have a much more complicated boundary. Recall that this boundary is the Julia set, J_c. If we magnify portions of the boundary of any of the images

(a) $c = -1$ (b) $c = 0.3 - 0.4i$

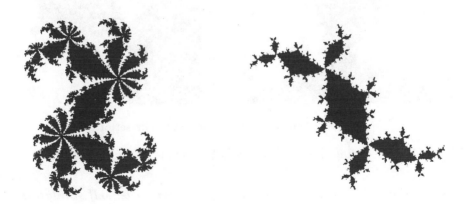

(c) $c = 0.360284 + 0.100376i$ (d) $c = -0.1 + 0.8i$

Fig. 16.8 Filled Julia sets for Q_c

in Figure 16.8, we see a definite pattern of self-similarity. For example, in Figure 16.9, we have magnified several portions of K_{-1} in succession. At each magnification, additional "decorations" attached to the filled Julia set are visible. This same phenomenon occurs when the other filled Julia sets in

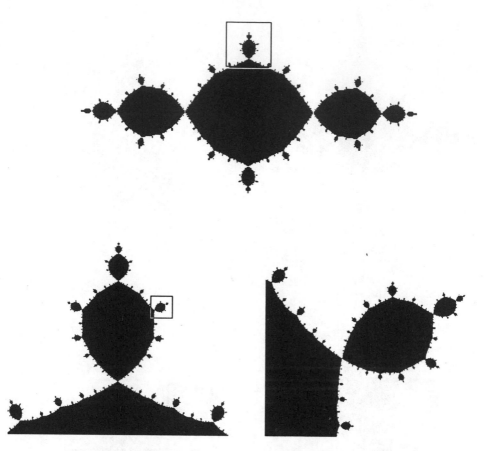

Fig. 16.9 Magnifications of the filled Julia set of Q_{-1}.

Figure 16.8 are magnified. This leads to our second observation.

Observation 2: *The Julia sets for Q_c appear to be self-similar sets reminiscent of fractals.*

For many c-values, K_c seems to be one connected albeit complicated shape. For many other c-values, however, the set K_c seems to consist of, at most, an isolated collection of points. In particular, if we compute K_c for a c-value with $|c| > 2$, we see this phenomenon. On the other hand, from our results in the previous section, we know that K_c is a Cantor set when $|c| > 2$. Thus this algorithm does not seem to work well in certain cases. We will remedy this in Section 16.6 when we present another algorithm to compute Julia sets.

Fig. 16.10 Filled Julia sets for $c = 0.25$ and $c = 0.255$.

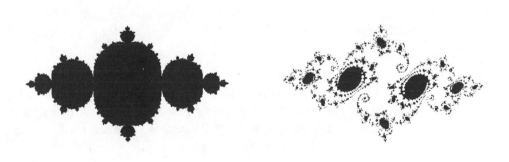

Fig. 16.11 Filled Julia sets for $c = -0.75$ and $c = -0.75 + 0.1i$.

We note, however, that the transition between large filled Julia sets and these sparse ones can be abrupt. For example, in Figure 16.10, we display the filled Julia sets for $c = 0.25$ and $c = 0.255$, each computed using 50 iterations. If you use 150 iterations to compute these images, you will see that the filled Julia set for $c = 0.25$ remains essentially unchanged, whereas the filled Julia set for $c = 0.255$ becomes decidedly smaller. Recall for later use that Q_c underwent a saddle-node bifurcation at $c = 0.25$. As another example, in Figure 16.11, we display the filled Julia sets for $c = -0.75$ and $c = -0.75 + 0.1i$, each computed using 100 iterations. Recall that Q_c underwent a period-doubling bifurcation at $c = -0.75$. Note how, in each case, the filled Julia sets seem to shatter into isolated pieces. We will explain this in the next chapter, but for now we simply note:

Observation 3: *For many c-values, K_c appears to consist of only a few isolated points or pieces. Also, K_c seems to change abruptly from one connected piece to many isolated points and pieces at certain special c-values.*

Remark. We emphasize that the pictures of the "shattered" Julia sets in Figures 16.10 and 16.11 are not quite correct. We will see in the next chapter that these filled Julia sets are actually Cantor sets. The large blobs seen in these images disappear when we use a larger number of iterations to compute these images.

16.5   Experiment: Filled Julia Sets and Critical Orbits

Goal: In this experiment you will investigate the relationship between the shape of the filled Julia set of Q_c and the fate of the orbit of 0 under iteration of Q_c.

Procedure: For each of the following c-values, use the computer to draw the filled Julia set of Q_c. Record whether the filled Julia set appears to be one connected piece or if it shatters into many isolated pieces as in Figures 16.10 and 16.11. Then compute the orbit of 0 for this c-value. Is this orbit bounded or unbounded? You should try at least the following c-values:

    a. $c = -1.5 + 0.2i$
    b. $c = -0.1 + 0.75i$
    c. $c = -0.4 + 0.8i$
    d. $c = 0.28 + 0.53i$
    e. $c = -0.11 + 0.86i$
    f. $c = -1.32$
    g. $c = 0.48 + 0.48i$
    h. $c = 1.5i$
    i. $c = -0.5 + 0.57i$
    j. $c = -0.4 + 0.4i$

Results: In a brief essay, describe the relationship between the shape of the filled Julia set and the fate of the orbit of 0.

Notes and Questions:

1. Many filled Julia sets consist of a collection of regions that are joined together at single points. For example, consider the filled Julia set of $Q_{-1}(z) = z^2 - 1$ displayed in Figure 16.9. Don't be fooled into thinking that these sets are disconnected. When in doubt, magnify portions of the set.

2. When computing the filled Julia set for Q_c, always begin with a square centered at the origin whose sides have length 4. By the Escape Criterion, orbits of points outside this square escape, unless $|c|$ happens to be larger than 2.

16.6   The Julia Set as a Repellor

In the first three sections of this chapter, we discussed three examples of specific Julia sets for Q_c. Each of these Julia sets exhibited certain properties:

   1. Repelling periodic points were dense in the Julia set.

   2. Q_c was supersensitive at any point in J_c.

Both of these properties hold for the Julia set of any Q_c. We cannot prove this here, since the proofs involve more advanced topics from complex analysis, specifically the theory of normal families. (See [Devaney, pp. 283–288] for details.) However, we will show that any repelling periodic point lies in J_c, as does any preimage of this point. This will lead us to a second computer algorithm that produces a picture of the Julia set of Q_c, but not the filled Julia set.

We begin with a standard result from complex analysis, whose proof demands somewhat more sophisticated techniques from analysis than previously encountered.

Cauchy's Estimate. *Suppose $P(z)$ is a complex polynomial. Suppose also that $|P(z)| \leq M$ for all z in the disk $|z - z_0| \leq r$. Then*

$$|P'(z_0)| < \frac{M}{r}.$$

Proof: If we first conjugate P via the translation $L(z) = z - z_0$, we may assume that $z_0 = 0$. Suppose then that

$$P(z) = \sum_{n=0}^{d} a_n z^n.$$

Consider the circle of radius r about 0 given by $t \mapsto re^{it}$. We have

$$\frac{1}{2\pi} \int_0^{2\pi} \frac{P(re^{it})}{re^{it}} dt = \frac{1}{2\pi} \int_0^{2\pi} \sum_{n=0}^{d} a_n r^{n-1} e^{i(n-1)t} dt$$

$$= \frac{1}{2\pi} \int_0^{2\pi} a_1 dt = a_1 = P'(0)$$

since the terms

$$\int_0^{2\pi} a_n r^{n-1} e^{i(n-1)t} dt$$

vanish when $n \neq 1$ as we see by integration of

$$\cos(n-1)t + i \sin(n-1)t$$

over the interval $0 \leq t \leq 2\pi$.

On the other hand,

$$\left| \frac{1}{2\pi} \int_0^{2\pi} \frac{P(re^{it})}{re^{it}} dt \right| \leq \frac{1}{2\pi} \int_0^{2\pi} \frac{|P(re^{it})|}{r} dt$$

$$\leq \frac{1}{2\pi} \int_0^{2\pi} \frac{M}{r} dt$$

$$= \frac{M}{r}.$$

Thus we have shown that

$$|P'(0)| \leq \frac{M}{r}.$$

This completes the proof.

Theorem. *Suppose z_0 is a repelling periodic point for Q_c. Then $z_0 \in J_c$.*

Proof: Suppose $z_0 \notin J_c$ but z_0 is a repelling periodic point with period n. Suppose $|(Q_c^n)'(z_0)| = \lambda > 1$. Then there is a disk centered at z_0 that lies in K_c and thus has the property that no orbit of a point in this disk escapes to infinity. Let r be the radius of this disk. Now, for each k, Q_c^{kn} is a polynomial. For each z in the disk we have, by the Escape Criterion,

$$|Q_c^{kn}(z)| \leq \max\{|c|, 2\}.$$

Let $M = \max\{|c|, 2\}$. By the Cauchy Estimate, for each k we must have

$$|(Q_c^{kn})'(z_0)| < \frac{M}{r}.$$

But

$$|(Q_c^{kn})'(z_0)| = \lambda^k \to \infty.$$

This contradiction establishes the result.

Corollary. *Suppose z_0 is a repelling periodic point for Q_c and z is a preimage of z_0. Then $z \in J_c$.*

The proof of this fact follows immediately from the previous theorem and the Boundary Mapping Principle.

We turn now to a second algorithm for computing Julia sets. Unlike the algorithm in the previous section, this one produces an image of the Julia set, not the filled Julia set. This algorithm rests on the fact that Q_c is supersensitive on J_c. Choose any $z \in \mathbf{C}$ (with at most one exception) and any point $z_0 \in J_c$. Let U be a neighborhood of z_0. By supersensitivity, there exists $w \in U$ and an iterate k such that $Q_c^k(w) = z$. This means that, given any $z \in \mathbf{C}$, we may find preimages of z under Q_c^k for large k arbitrarily close to any point in J_c.*

Thus, to find points in J_c, we will choose any $z \in \mathbf{C}$ and compute its "backward orbit." Of course, each point w with the exception of c has two preimages under Q_c, namely $\pm\sqrt{w - c}$. So, to compute the backward orbit of w, we will make a random choice of one of these two preimages at each stage.

Backward Iteration Algorithm: *Choose any point $z \in \mathbf{C}$. Compute 10,000 points on the backward orbit of z, randomly choosing one of the two possible preimages of Q_c at each stage. Plot all but the first 100 iterations.*

---

*   There is only one exception to this for any Q_c, namely the special point $z = 0$, which has only itself as a preimage for $Q_0(z) = z^2$. Hence there are no preimages of 0 under Q_0^k near the Julia set. See [Devaney, p. 275].

Some of the results of applying this algorithm are displayed in Figure 16.12. While this algorithm produces a fairly accurate picture of the Julia set, you will see that points in J_c that are reached by backward iteration are not evenly distributed in J_c. This is evident in Figure 16.12a, where there are small gaps in the curve that forms the Julia set of $Q_{-1}(z) = z^2 - 1$.

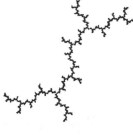

(a) $c = -1$ (b) $c = -i$

(c) $c = -0.8 + 0.3i$ (d) $c = 0.4 + 0.07i$

Fig. 16.12 Julia sets computed via backward iteration.

Remarks:

1. This algorithm, unlike the algorithm for the filled Julia set, works particularly well when the Julia set is a Cantor set. In Figure 16.13, we have displayed the Julia set of Q_c and several magnifications when $c = 0.5$. Note how more pieces of the Julia set become visible as we zoom in. As we will see in the next chapter, this Julia set is in fact a Cantor set, even though it looks quite different from the linear Cantor sets we studied in Chapter 7.

2. This algorithm is reminiscent of the iterated function systems we studied in Chapter 14. Instead of using a collection of linear contractions of the plane, here we randomly iterate a pair of nonlinear functions, namely $\pm\sqrt{w - c}$.

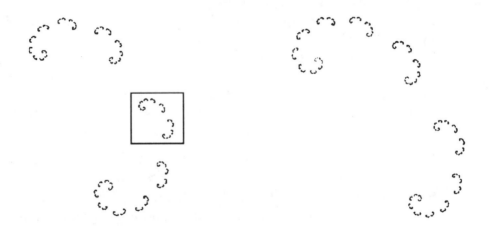

Fig. 16.13 The Julia set of $z^2 + 0.5$ and several magnifications.

Exercises

1. Describe the filled Julia set for $F(z) = z^3$.

2. Describe the filled Julia set for $F(z) = z^d$ for $d \geq 4$.

3. Consider $Q_2(z) = z^2 + 2$. Let $C = \{z \mid |z| = 2\}$. Give an accurate sketch of $Q_2^{-1}(C)$ and $Q_2^{-2}(C)$.

4. Use the techniques of Section 16.2 to conjugate $F(z) = z^3$ to a polynomial $P(z)$ via $H(z) = z + \frac{1}{z}$. What is $P(z)$?

5. The Saddle-Node Bifurcation.

    a. Find all complex c-values for which $Q_c(z) = z^2 + c$ has a fixed point z_0 with $Q_c'(z_0) = 1$.

    b. Consider the fixed points of Q_c in the complex plane for $c < 1/4$, $c = 1/4$, and $c > 1/4$. Determine whether these fixed points are attracting, repelling, or neutral. This is the complex saddle-node bifurcation.

    c. Determine the set of all complex c-values for which Q_c has an attracting fixed point. Sketch this set of c-values in the plane. Where does this set meet the real axis?

    d. Determine the set of complex c-values at which the function $F_c(z) = z^3 + c$ has a saddle-node bifurcation similar to that in part b.

6. The Period-Doubling Bifurcation.

    a. Find all complex c-values for which $Q_c(z) = z^2 + c$ has a fixed point z_0 with $Q_c'(z_0) = -1$.

    b. Show that the points

$$q_\pm(c) = -\frac{1}{2} \pm \frac{1}{2}\sqrt{-3 - 4c}$$

    lie on a 2-cycle unless $c = -3/4$.

    c. Determine whether this cycle is attracting, repelling, or neutral in the two real cases $-5/4 < c < -3/4$ and $c > -3/4$.

    d. Sketch the locations of the fixed points and $q_\pm$ for Q_c in the three real cases $-5/4 < c < -3/4$, $c = -3/4$, and $c > -3/4$. This is the complex period-doubling bifurcation.

7. Consider the complex function $G_\lambda(z) = \lambda(z - z^3)$. Show that the points

$$p_\pm(\lambda) = \pm\sqrt{\frac{\lambda + 1}{\lambda}}$$

lie on a cycle of period 2 unless $\lambda = 0$ or -1. Discuss the bifurcation that occurs at $\lambda = -1$ by sketching the relative positions of $p_\pm$ and the fixed points of G_λ as λ increases through -1, assuming only real values.

8. Let $Q_i(z) = z^2 + i$. Prove that the orbit of 0 is eventually periodic. Is this cycle attracting or repelling? Use the Backward Iteration Algorithm to compute the filled Julia set for Q_i. Does this set appear to be connected or disconnected?

9. The Logistic Functions. The following exercises deal with the family of logistic functions $F_\lambda(z) = \lambda z(1 - z)$ where both λ and z are complex numbers.

    a. Prove that $F_\lambda'(z) = \lambda(1 - 2z)$ using complex differentiation.

    b. Find all fixed points for F_λ.

    c. Find all parameter values λ for which F_λ has an attracting fixed point.

10. Prove that the complex function $H(z) = z + \frac{1}{z}$ takes straight rays emanating from the origin to hyperbolas as depicted in Figure 16.4.

CHAPTER 17

The Mandelbrot Set

Our goal in this chapter is to present some of the mathematics behind one of the most intricate and interesting shapes in dynamics, the Mandelbrot set. This set, first viewed in 1980 by the mathematician Benoit Mandelbrot, has been the subject of intense research ever since.

17.1   The Fundamental Dichotomy

In the previous chapter we described several filled Julia sets rigorously and many others experimentally. We noted that the filled Julia sets seemed to fall into one of two classes: those that were connected and those that were totally disconnected. There was apparently no "in between": either K_c consists of one connected piece or infinitely many pieces. Here we will show that this is indeed the case. In fact, this is one of the fundamental results of complex dynamics:

The Fundamental Dichotomy. *Let $Q_c(z) = z^2 + c$. Then either*
  1. *The orbit of the critical point 0 escapes to infinity, in which case K_c consists of infinitely many disjoint components, or*
  2. *The orbit of 0 remains bounded, in which case K_c is connected.*

Just as we discussed in Chapter 16, in the first case it can be proved that K_c is a Cantor set and the dynamics of Q_c on K_c is conjugate to the shift. Hence we really have a stronger statement: either K_c is totally disconnected or K_c is connected.

We have already shown that if $|c| > 2$, then the orbit of 0 escapes to infinity and K_c consists of infinitely many components. Hence we consider here only the case where $|c| \leq 2$. Note that there is good reason for separating these two cases. We have seen that K_{-2} is the interval $[-2, 2]$, which is connected, and that the orbit of 0 under Q_{-2} is eventually fixed. Hence $|c| = 2$ is the exact largest value for which K_c may be connected.

Recall that the escape criterion gives us a sufficient condition for an orbit to escape when $|c| \leq 2$: if $|Q_c^n(0)| > 2$ for some n, then the orbit of 0 tends to infinity. To prove the Fundamental Dichotomy, let's first suppose that $|Q_c^n(0)| \rightarrow \infty$ as $n \rightarrow \infty$ for some c with $|c| \leq 2$. Then there is a first iteration, say k, for which $|Q_c^k(0)| > 2$. Let $\rho = |Q_c^k(0)|$ and consider the circle of radius ρ, C_ρ, centered at the origin. The quadratic map Q_c maps points on this circle to the circle of radius ρ^2 centered at c in the plane. However, by Corollary 3 of the escape criterion, we know that the orbits of all points on C_ρ escape to infinity, and, moreover, the image circle lies in the exterior of the disk $|z| \leq \rho$. Let $D_\rho = \{z \, | \, |z| \leq \rho\}$. The circle $Q_c(C_\rho)$ actually encircles C_ρ as shown in Figure 17.1 since the center of this circle, c, satisfies $|c| \leq 2 < \rho$. By the Boundary Mapping Principle, the interior of the disk D_ρ is mapped to the interior of $Q_c(D_\rho)$. We therefore have a situation similar to that of the Cantor set case discussed in Section 16.3.

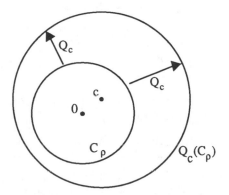

Fig. 17.1 The case where the orbit of 0 escapes.

Now consider $Q_c^{-1}(D_\rho)$. The set $Q_c^{-1}(D_\rho)$ is completely contained in the interior of D_ρ since C_ρ is mapped outside D_ρ. Moreover, 0 belongs to $Q_c^{-1}(D_\rho)$ since $Q_c(0) = c$ and $|c| < \rho$. Consequently, $Q_c^{-1}(D_\rho)$ is bounded by a simple closed curve.

Since k is the smallest integer for which $Q_c^k(0) \in C_\rho$, it follows that we may repeat the above argument $k - 1$ times. We find that

$$Q_c^{-(k-1)}(D_\rho) \subset \cdots \subset Q_c^{-1}(D_\rho) \subset D_\rho$$

with $Q_c^{-j}(D_\rho)$ contained in the interior of $Q_c^{-j+1}(D_\rho)$ for $j = 1, \ldots, k - 1$. Moreover, $Q_c^{-j}(D_\rho)$ is bounded by a simple closed curve which is mapped by Q_c to the boundary of $Q_c^{-j+1}(D_\rho)$, as shown in Figure 17.2.

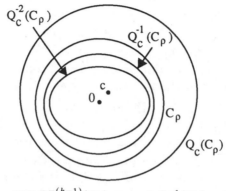

Fig. 17.2 $Q_c^{-(k-1)}(D_\rho) \subset \cdots \subset Q_c^{-1}(D_\rho) \subset D_\rho$.

At the kth iteration, this picture changes. The critical point 0 now belongs to $Q_c^{-k}(D_\rho)$. Since $Q_c^k(0) \in C_\rho$, it follows that 0 lies on the boundary of $Q_c^{-k}(D_\rho)$. Hence the boundary of $Q_c^{-k}(D_\rho)$ is a figure-eight curve, and each of the two lobes in the interior of this curve are mapped in one-to-one fashion by Q_c onto the interior of $Q_c^{-k+1}(D_\rho)$.

Thus we see that the preimage of D_ρ pinches into two lobes precisely when 0 lies on the boundary of $Q_c^{-k}(D_\rho)$.

The remainder of the argument is similar to that in Section 16.3. The set $Q_c^{-k-1}(D_\rho)$ consists of a pair of figure-eights and their lobes lying in the interior of $Q_c^{-k}(D_\rho)$. In general, $Q_c^{-k-n}(D_\rho)$ consists of 2^n disjoint figure-eights and their lobes, so we see that the filled Julia set decomposes into infinitely many disjoint components when the orbit of 0 escapes. This proves Part 1 of the Fundamental Dichotomy.

The only other possibility is $Q_c^k(0)$ never lies in the exterior of the circle of radius 2. In this case, for any $\rho > 2$, $Q_c^{-k}(D_\rho)$ is always a connected region bounded by a simple closed curve. By the Connectedness Property (Appendix A.3), $\cap_{n \geq 0} Q_c^{-n}(D_\rho)$ is a connected set. As usual, this is the filled Julia set. This completes the proof.

We must emphasize the incredible power and elegance of this result. For quadratic polynomials, the filled Julia set can assume one of only two possible shapes: a Cantor set or a connected set. There are no filled Julia sets that consist of 2 or 40 or 100 pieces. Either the filled Julia set consists of one piece or else it shatters into infinitely many distinct pieces. Moreover, and this is a truly amazing fact, it is the orbit of the critical point that determines which case holds. The orbit of this one special point determines the global structure of the set of all bounded orbits.

17.2   The Mandelbrot Set

The Fundamental Dichotomy indicates that there are only two basic types of filled Julia sets for Q_c, those that are connected and those that consist of infinitely many disjoint components. Moreover, it is the orbit of 0 that determines which of these two cases hold. This leads to an important definition.

Definition. The *Mandelbrot set* $\mathcal{M}$ consists of all c-values for which the filled Julia set K_c is connected. Equivalently,

$$\mathcal{M} = \{c \in \mathbf{C} \,|\, |Q_c^n(0)| \not\to \infty\}.$$

It is important to realize that $\mathcal{M}$ is a subset of the c-plane, not the dynamical plane where the Julia sets live. The Mandelbrot set is a subset of the parameter space for quadratic polynomials.

The escape criterion of Chapter 16 also gives us an algorithm for computing the Mandelbrot set. As before, we consider a rectangular grid in the c-plane. As we know, $\mathcal{M}$ is contained in the circle of radius 2 centered at the origin, so we will always assume that our grid is contained inside the square centered at the origin whose sides have length 4. For each point c in this grid, we compute the corresponding orbit of 0 under Q_c and ask whether or not this orbit tends to infinity. If the orbit does not escape, then our original point is in $\mathcal{M}$, so we will color the original point black. If the orbit escapes, then we leave the original point white.

Algorithm for the Mandelbrot Set: *Choose a maximum number of iterations, N. For each point c in a grid, compute the first N points on the*

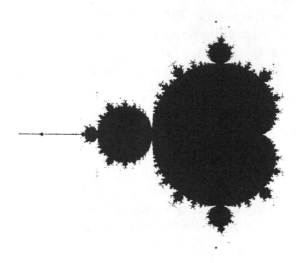

Fig. 17.3 The Mandelbrot set.

orbit of 0 under Q_c. If $|Q_c^i(0)| > 2$ for some $i \leq N$, then stop iterating and color c white. If $|Q_c^i(0)| \leq 2$ for all $i \leq N$, then color c black.

Remarks:

1. As with our algorithm for displaying filled Julia sets, this algorithm is not foolproof. There may be points for which the orbit of 0 takes more than N iterations to escape from the circle of radius 2. These points will be colored black even though they do not lie in $\mathcal{M}$. Despite this, it is advisable to keep the number of iterations low when computing $\mathcal{M}$, since the computations involved may take hours on a personal computer.

2. For the colored plates illustrating the Mandelbrot set, we used the same coloring scheme as we used for the filled Julia sets, namely, red c-values have critical orbits that escape fastest, followed by orange, yellow, green, blue, indigo, and violet.

We have included a program that effects this algorithm in Appendix B. As before, we encourage you to experiment by viewing different regions of $\mathcal{M}$ (Fig. 17.3). You will quickly see that the Mandelbrot set is one of the most fascinating objects in Mathematics!

Let's try to understand what this image means. Suppose Q_c has an attracting periodic point of some period. Then the entire basin of attraction of this point lies in K_c, and so the filled Julia set is not a Cantor set. By our discussion earlier, this means that c should lie in the Mandelbrot set. The question is where?

To answer this, let's first determine the set of c-values for which Q_c has an attracting or a neutral fixed point. We denote this set by C_1 and again emphasize that this is a subset of the c-plane. Let z_c denote the corresponding fixed point for Q_c when $c \in C_1$. Then z_c must satisfy the following pair of equations:

$$Q_c(z) = z^2 + c = z$$
$$|Q_c'(z)| = |2z| \leq 1.$$

The second equation says that z_c lies on or inside the disk bounded by the circle of radius $1/2$. On this circle, $|Q_c'(z_c)| = 1$, so z_c is a neutral fixed point. If we write this circle as $z = \frac{1}{2}e^{i\theta}$, then we find from the first equation that the c-values for which Q_c has a neutral fixed point lie on the curve given by

$$c = \zeta(\theta) = \frac{1}{2}e^{i\theta} - \frac{1}{4}e^{2i\theta}.$$

This curve is the cardioid depicted in Figure 17.4. It gives the boundary of the large black cardioid region visible in Figure 17.3. The interior of this region is the set of c-values for which Q_c has an attracting fixed point.

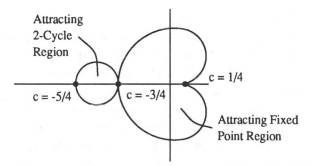

Fig. 17.4 The regions C_1 and C_2

Note that the boundary of C_1 meets the real axis at two points, $c = \frac{1}{4}$, the saddle-node bifurcation point, and $c = -\frac{3}{4}$, the period-doubling bifurcation point, or when $\theta = 0$ and π, respectively. Note also that if $c = \zeta(\theta)$ lies on this curve, then $Q'_c(z_c) = e^{i\theta}$. This means that, as the parameter θ winds once around the cardioid, the corresponding derivative at the fixed point $Q'_c(z_c)$ winds once around the unit circle. Let's summarize these facts in a Proposition.

Proposition (The Main Cardioid). *The region C_1 is bounded by a cardioid given by*

$$\zeta(\theta) = \frac{1}{2}e^{i\theta} - \frac{1}{4}e^{2i\theta}.$$

For c-values in the interior of this region, Q_c has an attracting fixed point. For c-values on the boundary of this region, Q_c has a neutral fixed point. Moreover, as θ increases from 0 to 2π, the derivative $Q'_{\zeta(\theta)}$ at the neutral fixed point winds once around the unit circle.

Now let's turn to the region C_2 in which Q_c has an attracting or neutral cycle of period 2. This region can also be computed explicitly, but we will leave the details as an exercise. The relevant results are summarized in the following Proposition.

Proposition (The Period-2 Bulb). *The region C_2 is bounded by a circle given by $|c + 1| = \frac{1}{4}$, the circle of radius $\frac{1}{4}$ centered at -1. Moreover, as c traverses this circle, the derivative $(Q_c^2)'$ at the 2-cycle winds once around the unit circle.*

The proof of this result is not difficult. See Exercises 1 and 2. This result is also illustrated in Figure 17.4. Note that C_2 meets the real axis in the c-plane at the period-doubling points $c = -3/4$ and $-5/4$. In the Mandelbrot set, the period-2 bulb is the large circular region visible just to the left of the main cardioid. The previous two propositions suggest that each of the decorations consist of c-values for which the corresponding quadratic functions have an attracting cycle of a given period. Further evidence of the truth of this fact is provided by the orbit diagram for Q_c. In Figure 17.5, we have superimposed the orbit diagram and the Mandelbrot set so that corresponding real c-values are aligned. That is, the real c-values that form the horizontal axis in the orbit diagram lie directly below the corresponding real c-values in the Mandelbrot set. Note that the real c-values for which we have an attracting fixed point or 2-cycle match up exactly, as we know to be the case by the previous two propositions. Similarly, the smaller sequence of bulbs along the real axis match up with the period-doubling regime in the orbit diagram. Finally, note the small "blip" along the tail of the Mandelbrot set above the period-3 window. We have magnified this region in Figure 17.6 and superimposed it on top of the period-3 window. Again we see that each decoration on the Mandelbrot set corresponds to a region (at least when c is real) where Q_c has an attracting cycle of a given period.

As we have seen, the orbit of 0 determines completely the connectivity of the filled Julia set of Q_c. But there is much more to this story. As we saw in Chapter 12, for real c, if Q_c admits an attracting periodic orbit, then the orbit of 0 must be attracted to it. The same fact is true for complex quadratic polynomials, but the proof demands advanced techniques that are beyond the scope of this book. (See [Devaney, p. 281] for a discussion of this.) Note that the Schwarzian derivative of a complex function may be complex, so the ideas we exploited when the Schwarzian derivative was negative do not hold in this case.

It thus follows that Q_c may have at most one attracting cycle for each c, and the orbit of 0 must find this cycle if it exists. This gives another important fact about $\mathcal{M}$: the interior of each decoration or bulb attached to $\mathcal{M}$ consists of c-values for which the corresponding Q_c has an attracting cycle of some given period. Again we will not prove this here, but instead will allow you to discover this yourself as part of the following experiments.

17.3   Experiment: Periods of Other Bulbs

Goal: In the previous section, we have seen some evidence that each decoration or bulb in the Mandelbrot set corresponds to a set of c-values for which Q_c has an attracting cycle of some given period. In this experiment you are to investigate this phenomenon further. We will try to determine an algorithm that allows us to specify the periods of all bulbs that hang off the main cardioid in $\mathcal{M}$.

Procedure: First use the computer to draw a picture of $\mathcal{M}$. Using a mouse or some other means of selecting coordinates from the screen, choose c-values from a variety of different bulbs that are tangent to the main cardioid in $\mathcal{M}$. For each chosen c, compute the orbit of 0 under Q_c and observe the ultimate behavior. This orbit should be attracted to an attracting periodic orbit of some period. Record both the c-value and the period of this cycle. Now choose a second c within the same bulb. Does the period change? Note: it is important to choose c-values near the center of the bulbs, for convergence

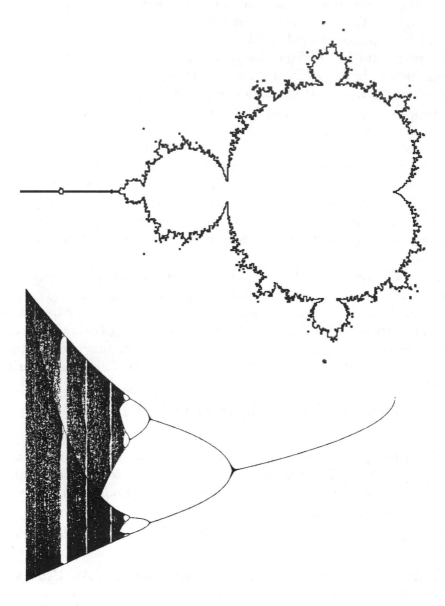

Fig. 17.5 The orbit diagram and $\mathcal{M}$ aligned vertically.

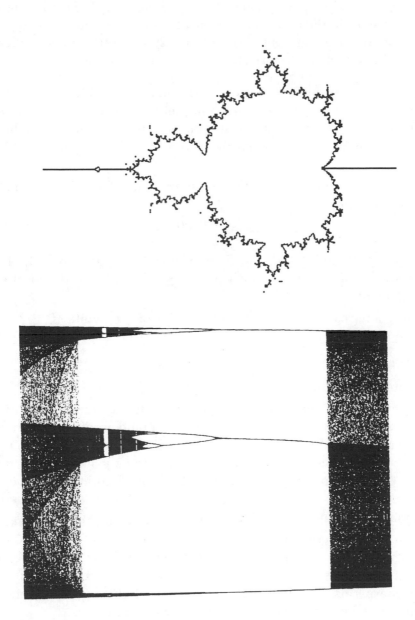

Fig. 17.6 The period 3 window and the corresponding portion of $\mathcal{M}$.

to the attracting cycle is very slow near the boundaries. This is related to the smears we see in the orbit diagrams near period doublings. Finally, draw a picture of the main cardioid with at least 20 decorations and indicate in each bulb the period of the corresponding attracting cycle.

Results: Given two bulbs attached to the main cardioid, you can usually distinguish with your eye the largest one between them. What is the corresponding period of this bulb? Can you determine a rule for finding the period of the largest bulb between two given bulbs? Describe this rule in a brief essay.

Now draw the unit circle in the complex plane. On this circle indicate all points of the form

$$z = e^{2\pi i (p/q)}$$

where $1 \le q \le 10$ and $0 \le p < q$. Then use the computer to find all bulbs attached to the main cardioid with period ≤ 10. Indicate the locations of these bulbs on the cardioid.

Is there any relationship between the order of the points $e^{2\pi i (p/q)}$ on the unit circle and the periods of the bulbs attached to the cardioid? Explain in an essay.

Notes and Questions:

1. One fact that will considerably shorten the time it takes to perform this and some subsequent experiments is that the Mandelbrot set is symmetric about the real axis. Moreover, the periods of symmetrically located bulbs are the same. We ask you to verify this in exercise 3 at the end of this chapter. Thus, for this experiment, you need only check the decorations in the upper half-plane.

2. Consider the small copy of the Mandelbrot set along the real c-axis that corresponds to the period-3 window in the orbit diagram. Use the computer to magnify this set. Using the same techniques as above, investigate the periods of the bulbs attached to the cardioid. Is there any relationship between the periods of these bulbs and those of the corresponding bulbs attached to the main cardioid? Explain in a brief essay.

3. Now investigate the periods of the bulbs attached to the large period-2 bulb just to the left of the main cardioid. Again, do you see any pattern? Explain in a brief essay.

17.4   Experiment: Periods of the Decorations

Goal: In this experiment you will hunt for all of the decorations in the Mandelbrot set that correspond to attracting cycles of period n. As you perform this experiment, you will undoubtedly view some of the wonderfully intricate geometric shapes that abound in $\mathcal{M}$.

Procedure: Use the computer to select a c-value from a decoration and then determine the period of the corresponding attracting cycle. Find all decorations that feature an attracting n-cycle for $n = 3$, 4, and 5.

Results: Indicate the locations of the period 3, 4, and 5 decorations on a copy of the Mandelbrot set. Some of these locations are indicated in Figure 17.7.

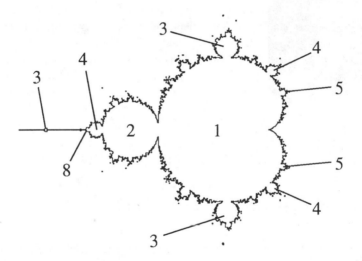

Fig. 17.7 Periods of some of the decorations in $\mathcal{M}$.

Notes and Questions:

1. There are exactly three period-3 decorations and six period-4 decorations.

2. Try to find all of the period-5 decorations. There are 15 of them. It is very difficult to find all of them, but it is fun to try!

3. Now try period 6, if you are a glutton for punishment. There are 27 of them! Good luck! While it may seem that these decorations are ordered in no apparent fashion around $\mathcal{M}$, there is in fact a beautiful description of how these bulbs are arranged due to Douady and Hubbard in 1983. Unfor-

Adrien Douady and John Hubbard

In the early 1980's, *Adrien Douady* (1935–) and *John Hubbard* (1945–) joined forces to study the structure of the Mandelbrot set. Using tools from complex analysis and dynamical systems, they managed to prove that, despite appearances, this set is connected. Moreover, using their celebrated technique involving "external rays," they were able to describe the dynamical meaning of the decorations attached to the Mandelbrot set as well as the dynamics on the corresponding Julia sets. All of the experimental results you will obtain in this chapter can be proved rigorously using these techniques.

At this time, there is one outstanding conjecture about the Mandelbrot set: that the set has the topological property of being *locally connected*. If the Mandelbrot set is locally connected, then the results of Douady and Hubbard imply that we understand more or less completely the dynamical behavior of quadratic polynomials.

Currently, Douady is a Professor of Mathematics at the École Normale Superieure in Paris and Hubbard is a Professor of Mathematics at Cornell University.

tunately, complete details of their work are too advanced for this text. See the forthcoming book of Milnor* for more details.

17.5   Experiment: Find the Julia Set

Goal: The aim of this experiment is to show you how each decoration in the Mandelbrot set corresponds to c-values whose Julia sets are qualitatively similar.

Procedure: Figure 17.8 depicts eight different filled Julia sets. Your task is to identify the decoration in the Mandelbrot set that contains the c-value that produced these Julia sets. In each case, use the computer to select c-values from a decoration in the Mandelbrot set and then draw the corresponding Julia set. If your computer is slow, it might help to use the backwards iteration algorithm to produce this image.

---

*   J. Milnor, *Dynamics in One Complex Variable: Introductory Lectures* (SUNY Stony Brook, Institute for Mathematical Sciences, Preprint #1990/5, 1990).

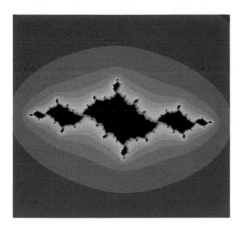

Fig. 17.8a

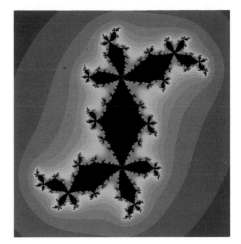

Fig. 17.8b

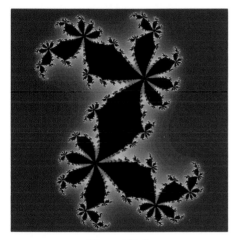

Fig. 17.8c

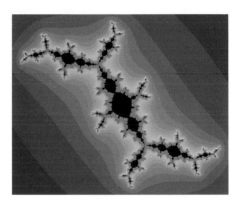

Fig. 17.8d

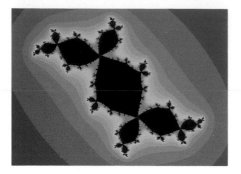

Fig. 17.8e

Fig. 17.8f

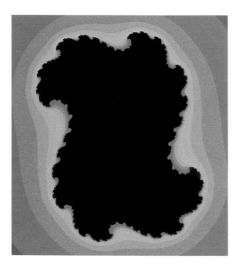

Fig. 17.8g

Fig. 17.8h

Fig. 17.9a   Period 4 bulb

Fig. 17.9b   Period 5 bulb

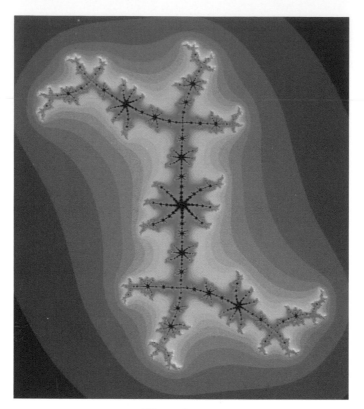

Fig. 17.10a.

Julia set corresponding to junction point in Fig. 17.9a

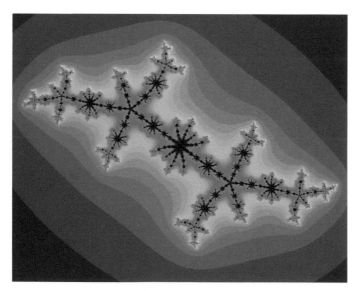

Fig. 17.10b.

Julia set corresponding to junction point in Fig. 17.9b

Results: On an image of the Mandelbrot set, indicate the approximate locations of Julia sets "a" through "h" in Figure 17.8.

Notes and Questions:

1. As a hint, each of the Julia sets (with only one exception) come from decorations directly attached to the main cardioid. You may have to hunt within a given decoration for the exact Julia set depicted, but you will quickly see that all filled Julia sets from a single decoration have the same essential features, that is, are qualitatively the same.

2. Do you recognize any connection between the structure of the Julia sets from a given decoration and the period of that decoration, as described in the previous experiments?

3. Once you have found the locations of each of these Julia sets, investigate the shapes of the Julia sets in corresponding locations in smaller copies of the Mandelbrot set, particularly the period-3 Mandelbrot set on the real axis. Is there any relation between these images? Explain in a brief essay.

17.6   Experiment: Spokes and Antennas

Goal: In this experiment we investigate the relationship between the periods of bulbs in the Mandelbrot set and the shapes of the antennas that protrude from these decorations.

Procedure: Use the computer to magnify a variety of different bulbs attached to the main cardioid of $\mathcal{M}$. Be sure to include both the bulb and its attached antenna as illustrated in Figure 17.9. For each decoration, record the period of the bulb attached to the main cardioid. Also, count and record the number of "spokes" emanating from the central junction point of the attached antenna.

Results: Is there a relationship between the period of the bulbs attached to the main cardioid and the number of spokes in its antenna? Explain in a brief essay.

Notes and Questions:

1. Investigate the antennas and decorations attached to other bulbs in $\mathcal{M}$. Is there a relation between the number of spokes and the periods? Explain your findings in a brief essay.

2. Compare the shapes of the antennas attached to corresponding bulbs hanging off the main cardioid, the circular period 2 bulb, and the small period-3 Mandelbrot set along the real axis. Are there differences among

these similarly placed antennas? Are they longer or shorter? How do they differ from one another?

17.7   Experiment: Similarity of the Mandelbrot Set and Julia Sets

Goal: Our goal in this experiment is to view a remarkable relationship between the structure of the Mandelbrot set near special c-values and the shape of the corresponding filled Julia sets.

Procedure: For a variety of different decorations attached to the main cardioid, magnify the central portion of the antenna from which all of the spokes emanate. Zoom in on this point several times. Now, with a mouse or other means of selecting coordinates from the screen, determine as exactly as possible the corresponding c-value. Finally, use the computer to draw the filled Julia set for this c-value using backward iteration (Fig. 17.10).

Results: Compare the shape of the magnified Mandelbrot set near the junction point with the corresponding filled Julia set. Are there any similarities between these two images? Explain in an essay.

Notes and Questions: The similarity between magnified portions of the Mandelbrot set and the corresponding filled Julia sets holds only near certain c-values such as the central junctions of the antenna. These are c-values for which 0 is eventually periodic. Such c-values are called *Misiurewicz points*.

Exercises

1. Prove that Q_c has a periodic point of prime period 2 at each root of the equation $z^2 + z + c + 1 = 0$. *Hint:* Recall a similar result proved in the real case in Section 6.1.

2. Show that Q_c has an attracting 2-cycle inside the circle of radius $1/4$ centered at -1.

3. Prove that the Mandelbrot set is symmetric about the real axis. *Hint:* Show this by proving that Q_c is conjugate to $Q_{\bar{c}}$. Show that your conjugacy takes 0 to 0. Therefore the orbit of 0 has similar fates for both Q_c and $Q_{\bar{c}}$.

The Logistic Functions. The following seven exercises deal with the logistic family $F_\lambda(z) = \lambda z(1 - z)$ where both λ and z are complex numbers.

4. Prove that, for $\lambda \neq 0$, $|F_\lambda(z)| > |z|$ provided $|z| > \frac{1}{|\lambda|} + 1$. Use this to give the analog of the escape criterion for the logistic family.

5. Show that if $|\lambda| > 2 + \sqrt{2}$, the orbit of the critical point of F_λ escapes to infinity.

6. Show that, if $\lambda \neq 0$, the logistic function F_λ is conjugate to the complex quadratic function $Q_c(z) = z^2 + c$ where $c = \frac{\lambda}{2} - \frac{\lambda^2}{4}$. Let $c = V(\lambda)$ be this correspondence between λ and c. Why does this result fail if $\lambda = 0$?

7. Let L_1 be the set of λ-values for which F_λ has an attracting fixed point. Compute L_1 and sketch its image in the complex plane.

8. Show that the image of L_1 under the correspondence V in exercise 6 is the main cardioid in the Mandelbrot set.

9. Identify geometrically the set L_2 of λ-values that are mapped to the period-2 bulb in the Mandelbrot set by the correspondence V. How many connected regions are in this set?

10. Modify the algorithm for the Mandelbrot set to produce the analog of this set for the logistic family. Use the escape criterion discussed in exercise 4. After running this program, identify the regions L_1 and L_2 in this set.

Higher Degree Polynomials. The following eight exercises deal with polynomials of the form $P_{d,c}(z) = z^d + c$.

11. Prove that $P'_{d,c}(z) = dz^{d-1}$. Conclude that, for each integer $d > 1$, $P_{d,c}$ has a single critical point at 0.

12. Prove the following escape criterion for $P_{d,c}$. Show that if $|z| \geq |c|$ and $|z|^{d-1} > 2$, then $|P^n_{d,c}(z)| \to \infty$ as $n \to \infty$.

13. Show that if $|c| > 2^{\frac{1}{d-1}}$, the orbit of the critical point of $P_{d,c}$ escapes to infinity.

14. Suppose $|c| > 2^{\frac{1}{d-1}}$. In a brief essay, discuss the structure of the filled Julia set for $P_{d,c}$. What is the shape of the preimage of the circle of radius $|c|$ under $P_{d,c}$? Your discussion should parallel that of the discussion in Section 16.3.

15. Use the results of exercise 12 to modify the algorithm for the Mandelbrot set to produce the analog of this set for the family $P_{d,c}$. These sets are called the *degree-d bifurcation sets*. Two are displayed in Figure 17.11.

16. For a fixed value of d, find the set of c-values for which $P_{d,c}$ has an attracting fixed point. Find an expression for the boundary of this region. Identify this region in the images generated by the previous exercise.

Fig. 17.11 The degree-3 and degree-4 bifurcation sets.

17. Prove that the degree-3 bifurcation set is symmetric with respect to reflection through the origin. *Hint:* See exercise 3 above.

18. Prove that the degree-4 bifurcation set is symmetric with respect to rotation through angle $2\pi/3$. *Hint:* Show that Q_c is conjugate to Q_d if $d = e^{2\pi i/3}c$. Generalize this and the result of the previous exercise to the degree-d bifurcation set.

19. Show that $c = -2$ and $c = i$ are Misiurewicz points for Q_c, that is, 0 is eventually periodic for Q_c. On a sketch of the Mandelbrot set, locate these two c-values as accurately as possible.

20. For c and z complex, consider the functions

$$F_c(z) = c\left(z^2 + \frac{1}{z^2}\right).$$

   a. What are the critical points of F_c?

   b. Prove that the orbits of all critical points have the same fate.

   c. Show that

$$|z| + \frac{1}{|z|^3} > \frac{1}{|c|}$$

      is an escape criterion for F_c, that is, if any point of the orbit of z_0 satisfies this condition, then the orbit of z_0 escapes to infinity.

   d. Which orbit would you use to compute the analog of the Mandelbrot set for this family of functions?

CHAPTER 18

Further Projects and Experiments

In this chapter we suggest some further projects and experiments in dynamics that students may pursue. In many cases, these projects demand a combination of experimentation and rigorous mathematics. In each case, there are many phenomena that are, as yet, unexplained.

18.1   The Tricorn

Given a complex number $z = x + iy$, recall that its *complex conjugate* is defined as $\overline{z} = x - iy$. Note that $\overline{z}$ is obtained by simply reflecting z through the x-axis.

Iteration of the function $A_c(z) = \overline{z}^2 + c$ turns out to have ramifications for the study of higher degree polynomials. For example, although A_c is itself not a polynomial, its second iterate A_c^2 is the fourth-degree polynomial $A_c^2(z) = z^4 + 2\overline{c}z^2 + \overline{c}^2 + c$.

The critical point for A_c is 0, since $c = A_c(0)$ has only one preimage whereas any other $w \in \mathbf{C}$ has two preimages.

The Julia set and filled Julia sets for A_c are defined exactly as for the quadratic polynomials Q_c, and arguments similar to those in previous chapters show that the Julia set of A_c is either connected or totally disconnected, depending on whether the orbit of 0 is bounded or escapes to infinity. This suggests that we plot the analog of the Mandelbrot set for A_c. The resulting set is called the *tricorn*, since it resembles a tricornered hat. See Plates 35-37 in the tour in Chapter 1.

Project: First prove that if $|z| \geq \max(|c|, 2)$, then $A_c^n(z) \to \infty$. Then use this fact to modify the algorithm that plots the Mandelbrot set so that the tricorn is plotted.

Exercises

1. Prove that $A_0(z) = \bar{z}^2$ has four fixed points, not two as might be expected.

2. Show that $A_c^2(z)$ is a polynomial of degree 4. Is $A_c^3(z)$ a polynomial in z?

3. Since A_c is not a polynomial in z, we do not define its derivative with respect to z as we did in Section 15.4. However, since A_c^2 is a polynomial of degree 4, we may compute the derivative of A_c^2 as usual. Suppose z_0 is a fixed point for A_c. Show that $(A_c^2)'(z_0) = 4z_0\bar{z}_0$. Conclude that $(A_c^2)'(z_0)$ is always a real number.

4. First show that $\bar{z}^2 + c$ is conjugate to $\bar{z}^2 + d$, where $d = e^{2\pi i/3}$. Conclude that the tricorn is symmetric under rotations through angle $2\pi/3$.

5. Find an explicit formula for the curve in the c-plane along which A_c has a fixed point z_0 at which $(A_c^2)'(z_0) = 1$. This curve is called a *deltoid*.

18.2   Cubics

The study of cubic polynomials is much more complicated than the study of quadratics. The main reason is the typical cubic polynomial has two critical points, not just one. This means that we have several additional phenomena that may occur in this case. For example, a cubic polynomial may have two distinct attracting fixed or periodic orbits. Unlike quadratics, where we have two distinct cases (the critical orbit either escapes or is bounded), there are three possibilities for cubics: both critical orbits escape, both are bounded, and the new case, one critical orbit escapes and one remains bounded. Among other things, this means that the natural parameter space for cubics is four-dimensional, as there are two complex parameters. Hence plotting the full parameter space for cubics is much more difficult, and indeed this set is very poorly understood at present.

As in the case of quadratic polynomials, any cubic polynomial is conjugate to one of the special form

$$C_{a,b}(z) = z^3 + az + b$$

Bodil Branner

One of the leading experts in the study of the dynamics of cubic polynomials, *Bodil Branner* (1943–) has teamed with John Hubbard to produce the first extensive study of the analog of the Mandelbrot set for cubics*. Since there are two critical orbits for cubics, the parameter space for cubics is naturally four-dimensional. This means that the structure of this set is much more complicated than that of the Mandelbrot set. Nonetheless Branner's work indicates that a combination of techniques involving complex analysis and symbolic dynamics can shed light on the topology of this set. Branner is currently Associate Professor of Mathematics at the Technical University of Copenhagen.

where a and b are complex. We ask you to verify this in exercise 1. This clearly displays the two complex parameters necessary in the cubic case.

To plot filled Julia sets of cubics, we need a condition similar to the escape criterion of Chapter 16, which holds only for quadratic polynomials. One such condition is given by assuming

$$|z| > \max\left(|b|, \sqrt{|a|+2}\right).$$

Using the triangle inequality twice, we have

$$
\begin{aligned}
|C_{a,b}(z)| = |z^3 + az + b| \\
\geq |z|^3 - |az + b| \\
\geq |z|^3 - |a||z| - |b| \\
> |z|^3 - |a||z| - |z| \\
= |z|(|z|^2 - (|a|+1)) \\
> \lambda|z|
\end{aligned}
$$

for some $\lambda > 1$. As usual, we may apply this inequality n times to find

$$|C_{a,b}^n(z)| > \lambda^n|z|,$$

and hence the orbit of z must tend to infinity. We have proved:

---

*B. Branner and J. Hubbard, "The Iteration of Cubic Polynomials, Part I" *Acta. Math.* **66** (1988), pp. 143-206.

Escape Criterion for Cubics: *Let* $C_{a,b}(z) = z^3 + az + b.$ *Suppose*

$$|C^n_{a,b}(z)| > \max(|b|, \sqrt{|a| + 2})$$

for some n. Then the orbit of z escapes to infinity.

Project: Modify the program for filled Julia sets of quadratic polynomials so that the program computes the filled Julia set of the cubic $C_{a,b}(z) = z^3 + az + b$ where $a, b \in \mathbf{C}$.

Exercises and Experiments

1. Prove that any cubic polynomial is conjugate to one of the form $C_{a,b}(z) = z^3 + az + b$.

2. What are the critical points and critical values of $C_{a,b}$?

3. Use the program mentioned above to compute the filled Julia sets of $C_{a,b}$ where
    a. $a = -1, b = 1$
    b. $a = -1, b = 1.1$
Describe the bifurcation you see as b increases from 1 to 1.1 and a remains fixed at -1. What is the exact point of bifurcation? Analyze this bifurcation on the real line using graphical analysis.

4. Find values of a and b so that $C_{a,b}$ has the property that the critical points are fixed and distinct. Conclude that, for these values of a and b, $C_{a,b}$ has two distinct attracting fixed points. Describe the filled Julia set for $C_{a,b}$.

5. Find values of a and b so that both critical points lie on a cycle of period 2. Describe the filled Julia set for this function.

6. Consider the polynomial $P(z) = z^3 - 3z + 3$. Compute the critical points of P. Show that one critical point is fixed while the other has an orbit that tends to infinity. Use graphical analysis to describe the dynamics of this function on the real axis.

7. Describe the filled Julia set for the cubic polynomial in exercise 6.

8. Discuss the bifurcation that occurs in the family of real cubics $F_a(x) = x^3 - ax$ as a increases through 1. Then compute the filled Julia set of $C_{-a,0}(z) = z^3 - az$. What changes occur in the filled Julia set as a passes through the bifurcation value?

9. Find a family of cubic polynomials $C_{a,b}$ having the property that all three fixed points for $C_{a,b}$ tend to infinity as $|b| \to \infty$.

10. Consider the cubic polynomial $P(z) = z^3 - 1.6z + 1$. Use the algorithm above to compute the filled Julia set for this polynomial. Based upon what you see, what do think happens to the orbits of the two critical points of P? Use graphical analysis to explain this further.

11. Find an analog of the escape criterion for fourth-degree polynomials.

12. Prove that any fourth-degree polynomial is conjugate to one of the form

$$z^4 + az^2 + bz + c.$$

18.3   Exponential Functions

To define the complex exponential function, we need to recall one of the most beautiful formulas from elementary calculus, Euler's Formula. This formula relates the exponential and the trigonometric functions:

$$e^{ix} = \cos x + i \sin x.$$

We have used this formula before, but perhaps it is useful to recall why it is true. To do this, we simply use the power series representation for e^x evaluated at ix, finding

$$e^{ix} = \sum_{n=0}^{\infty} \frac{(ix)^n}{n!}$$

$$= 1 + ix - \frac{x^2}{2!} - i\frac{x^3}{3!} + \frac{x^4}{4!} + i\frac{x^5}{5!} + \cdots$$

$$= 1 - \frac{x^2}{2!} + \frac{x^4}{4!} \cdots + i(x - \frac{x^3}{3!} + \frac{x^5}{5!} \cdots)$$

$$= \cos x + i \sin x.$$

Note that these series converge absolutely for all x, so the above rearrangement is possible. Euler's formula allows us to define the complex exponential function.

Definition. The *complex exponential function* e^z is defined by

$$e^z = e^{x+iy} = e^x e^{iy}$$

$$= e^x \cos y + ie^x \sin y.$$

Note that e^{x+iy} is defined for all complex numbers. Furthermore the exponential is $2\pi i$-periodic, since

$$
\begin{aligned}
e^{x+iy+2\pi i} &= e^x \cos(y + 2\pi) + ie^x \sin(y + 2\pi) \\
&= e^x(\cos y + i \sin y) \\
&= e^{x+iy}.
\end{aligned}
$$

Furthermore, the complex exponential maps a vertical line of the form $x = c$ to a circle of radius e^c, since

$$
|e^{c+iy}| = |e^c(\cos y + i \sin y)| = e^c.
$$

Similarly, the exponential maps a horizontal line of the form $y = b$ to a ray emanating from the origin with angle $\theta = b$, since

$$
e^{x+ib} = e^x(\cos b + i \sin b).
$$

There is a fundamental difference in the filled Julia sets of maps like the exponential and our quadratic friends. No longer do we have an escape criterion. For example, consider the left half-plane given by

$$
H = \{x + iy \mid x \leq c\}.
$$

By our above remarks, any point in H is mapped by the exponential inside the circle of radius e^c centered at 0. Hence far away points in the left half-plane move very close to the origin after one iteration. Nonetheless, there is an "approximate" escape criterion that may be used to describe the set of orbits that escape. If the real part of z is large, say $z = x + iy$ with $x > 50$, then

$$
|e^z| = |e^x e^{iy}| = e^x > e^{50},
$$

which is very large. Although it is not true that any orbit that reaches the half-plane $x > 50$ eventually escapes, it can be shown that there is often a nearby point whose orbit does escape. A precise discussion of this point is beyond the scope of this book. For more details, we refer the reader to some recent work of Durkin*. Hence we use the following as our condition for escape:

---

* See Marilyn B. Durkin, "The Accuracy of Computer Algorithms in Dynamical Systems," *International Journal of Bifurcation and Chaos.* 1 No. 3, 1991.

Approximate Escape Criterion for Exponentials. *Suppose $E_\lambda(z) = \lambda e^z$. If the real part of $E_\lambda^n(z)$ exceeds 50, we say that the orbit of z "escapes."*

Remarks

1. It is known that, unlike the case for polynomials, any point whose orbit escapes for the complex exponential actually lies in the Julia set of $E_\lambda(z) = \lambda e^z$. This means that, arbitrarily close to such a point, there is a point on a repelling periodic orbit. Thus the structure of the Julia sets for functions like the complex exponential is much more complicated than that for polynomials. For a more complete discussion of this topic, we refer to [Devaney, pp. 319-326].

2. You will very quickly observe that computing filled Julia sets for exponentials is much more time consuming than those for polynomials. You may have to wait hours to compute one picture on a computer without a numeric coprocessor.

Project. Modify the program for filled Julia sets to sketch the escape and non-escape orbits for λe^z. Be sure to use the approximate escape criterion as your test for escape. As an example, the set of non-escape points for $(1 + 2i)e^z$ is displayed in black in Figure 18.1. White points in this picture correspond to points whose orbits eventually reach the region $\operatorname{Re} z > 50$.

Fig. 18.1 The filled Julia set for $(1 + 2i)e^z$.

Exercises and Experiments

1. Use the program mentioned above to describe the set of escaping orbits for e^z. Choose a large enough number of iterations so that the picture "stabilizes." What do you conclude about the set of escaping orbits?

2. Use the above program to describe the set of escaping orbits for $0.3e^z$. Is there a difference between this picture and that of exercise 1? What happens if you choose more iterations?

3. Compute an accurate graph of the real function $E_\lambda(x) = \lambda e^x$ for a selection of λ-values between 0.3 and 1. Discuss the dynamics of E_λ on the real line for these λ-values. Does a bifurcation occur? Explain.

4. Consider the function $E(x) = \frac{1}{e}e^x$. Prove that 1 is a neutral fixed point for E that weakly attracts from one side and weakly repels from the other.

5. Prove that there are no escaping orbits for $E(z) = \frac{1}{e}e^z$ in the half-plane $\{x + iy \mid x < 1\}$. *Hint:* What happens to the vertical line $x = 1$ under one iteration of E?

6. Find the set of complex λ-values such that $E_\lambda(z) = \lambda e^z$ has an attracting fixed point in the plane. Sketch this set in the λ-plane.

7. Use the program mentioned above to sketch the filled Julia sets for λe^z when λ is given by:
   a. πi
   b. $2\pi i$
   c. $-2.5 + 1.5i$
   d. $-2.7 + 1.5i$
   e. $1 + 0.2i$

8. Use an accurate graph and graphical analysis to discuss the dynamics of $E_\lambda(x) = \lambda e^x$ on the real line for a selection of λ-values in the range $-4 < \lambda < -2$. Do you see a bifurcation? Now compute the filled Julia sets for these values of λ. Describe what you see in an essay.

18.4 Trigonometric Functions

The complex trigonometric functions are defined to be linear combinations of the complex exponential as follows:

$$S_\lambda(z) = \lambda \sin z = \frac{\lambda}{2i}\left(e^{iz} - e^{-iz}\right)$$

$$C_\lambda(z) = \lambda \cos z = \frac{\lambda}{2} \left(e^{iz} + e^{-iz} \right).$$

The reason for these seemingly strange definitions is Euler's Formula. Recall that, for real θ-values, we have

$$e^{i\theta} = \cos\theta + i\sin\theta.$$

Hence

$$e^{-i\theta} = \cos\theta - i\sin\theta.$$

If we add these two equations, we find that

$$e^{i\theta} + e^{-i\theta} = 2\cos\theta,$$

which motivates our definition of the complex cosine. Subtracting the second equation from the first yields the complex sine.

As with the exponential, there is no "absolute" escape criterion. However, there is again an "approximate" test for escape. Suppose $\text{Im}\, z < -50$. Then, if $z = x + iy$,

$$|\sin z| = \frac{1}{2}|e^{iz} - e^{-iz}|$$

$$= \frac{1}{2}|e^{-y}e^{ix} - e^{y}e^{-ix}|$$

$$\geq \frac{1}{2}\left(|e^{-y}||e^{ix}| - |e^{y}||e^{-ix}|\right)$$

$$= \frac{1}{2}(e^{-y} - e^{y}).$$

Thus, if $\text{Im}\, z < -50$, the modulus of the image of z is quite large, since e^{-y} is large while e^{y} is small. Similar arguments work in the case where $\text{Im}\, z > 50$.

As in the case of the complex exponential, it is not true that points that satisfy $|\text{Im}\, z| > 50$ have orbits that necessarily tend to infinity, but it is known that nearby points have this property (at least for $\sin z$ and $\cos z$).

Approximate Escape Criterion for Trigonometric Functions. *Let $S_\lambda(z) = \lambda \sin z$. If $|\text{Im}\, z| > 50$, then we say that the orbit of z "escapes."**

---

*   A BASIC program that implements the approximate escape criterion for trigonometric functions may be found in Robert L. Devaney, *Chaos, Fractals, and Dynamics: Computer Experiments in Mathematics* (Menlo Park, CA: Addison-Wesley, 1990).

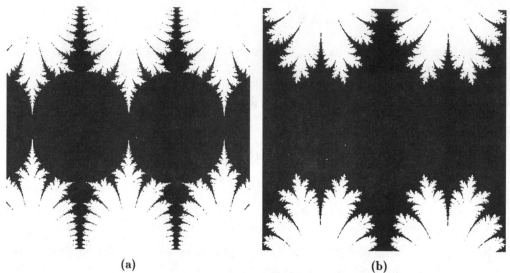

(a)   (b)

Fig. 18.2 The filled Julia sets for (a) $\sin z$, and (b) $\cos z$.

A similar criterion holds for $C_\lambda(z) = \lambda \cos z$. As with the complex exponential, it is true that the escaping points also lie in the Julia sets of S_λ or C_λ. The filled Julia sets for sine and cosine are shown in Figure 18.2.

Project: Modify the program for filled Julia sets to sketch the filled Julia set and escaping orbits for S_λ and C_λ.

Exercises and Experiments

1. Use the program mentioned above to compute the filled Julia set of $S_\lambda(z) = \lambda \sin z$ where $\lambda = 0.9$ and $\lambda = 1.1$. Describe the changes you see.

2. What are the dynamics of S_1 on the real line? Explain using graphical analysis.

3. Discuss the bifurcation that occurs when $\lambda = 1$ for S_λ with λ varying along the real line.

4. Show that, if y is a real number, then $S_\lambda(iy) = i\lambda \sin y$. Conclude that, if both y and λ are real, S_λ preserves the imaginary axis.

5. Discuss the dynamics of S_λ on the imaginary axis when λ is real and positive. What kind of bifurcation occurs when $\lambda = 1$?

6. Discuss the dynamics of S_λ on the imaginary axis when λ is real and negative. What kind of bifurcation occurs when $\lambda = -1$?

7. Use the computer to draw the filled Julia sets of C_λ when $\lambda - 0.6i$ and $\lambda = 0.7i$. Is there a difference?

8. Consider the complex cosines $C_{i\mu}(z) = i\mu \cos z$ where μ is a real parameter. Show that $C_{i\mu}$ preserves the imaginary axis. Use graphical analysis to discuss the bifurcation that occurs on this axis as μ passes through 0.67 ....

18.5 Complex Newton's Method

Recall from Chapter 13 that Newton's method is an iterative procedure to solve certain equations. Given a function F, to find the solutions of $F(x) = 0$, we simply choose an initial guess x_0 and then compute the orbit of x_0 under the Newton iteration function

$$N(x) = x - \frac{F(x)}{F'(x)}.$$

We saw earlier that roots of F were in one-to-one correspondence with attracting fixed points of N. Often, but not always, the orbit of x_0 converges to an attracting fixed point of N, which thus yields a root of F.

In this section we extend the discussion of Newton's method to the case where F is a complex polynomial. The same techniques used in Chapter 13 show that the complex Newton iteration function

$$N(z) = z - \frac{F(z)}{F'(z)}$$

has an attracting fixed point at z_0 if and only if z_0 is a solution of the equation $F(z) = 0$. Here z is a complex number. Hence we may try to use iteration of N to find roots of F in the complex plane.

Let's now redo some of the examples we worked out in Chapter 13, this time in the plane.

Example. Suppose $F(z) = z^2 - 1$ with roots at $z = \pm 1$. Recall from Chapter 13 that any real number $x_0 \neq 0$ has an orbit that tends to ± 1 under the real Newton iteration function.

In the complex case, the Newton iteration function is

$$N(z) = z - \frac{z^2 - 1}{2z} = \frac{1}{2}\left(z + \frac{1}{z}\right).$$

We first claim that all of the orbits of N tend to one of the roots, except those whose seeds lie on the imaginary axis. To see this, we construct a conjugacy with another old friend. Let

$$H(z) = \frac{z-1}{z+1}.$$

Note that H is defined except at $z = -1$ and that $H(-1) = \infty$. Now consider the diagram

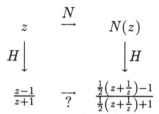

As usual, we ask which function completes this diagram? To answer this, note that

$$H \circ N(z) = \frac{\frac{1}{2}\left(z + \frac{1}{z}\right) - 1}{\frac{1}{2}\left(z + \frac{1}{z}\right) + 1} = \frac{z^2 - 2z + 1}{z^2 + 2z + 1} = \left(\frac{z-1}{z+1}\right)^2.$$

Thus, if $Q(z) = z^2$ is the complex squaring function, then

$$H \circ N(z) = Q \circ H(z).$$

Thus N is conjugate to Q! (It is easy to check that H is one-to-one; see exercise 1.)

Note that H takes the imaginary axis to the unit circle. This follows since

$$|H(iy)| = \left|\frac{iy - 1}{iy + 1}\right| = 1.$$

The quotient is equal to 1 since the complex numbers $iy - 1$ and $iy + 1$ are equidistant from the origin. Furthermore, H takes points z_0 with $\mathrm{Re}\, z_0 > 0$ to the interior of the unit circle and points with $\mathrm{Re}\, z_0 < 0$ to the exterior of the unit circle. In particular, H takes 1 to zero and -1 to infinity. So, by the conjugacy, if $\mathrm{Re}\, z_0 > 0$, then the orbit of z_0 tends to 1 under N, whereas if $\mathrm{Re}\, z_0 < 0$, the orbit of z_0 tends to -1.

Finally, if $\mathrm{Re}\, z_0 = 0$, H takes z_0 to the unit circle where Q is known to be chaotic (see Section 10.2). Thus we see that Newton's method for

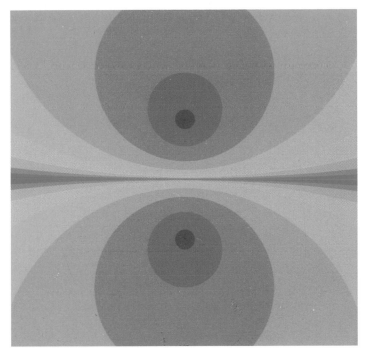

Fig. 18.4a.   Newton's method for $f(z) = z^2 + 1$

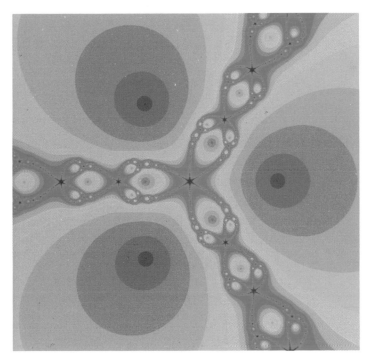

Fig. 18.4b.   Newton's method for $f(z) = z^3 - 1$

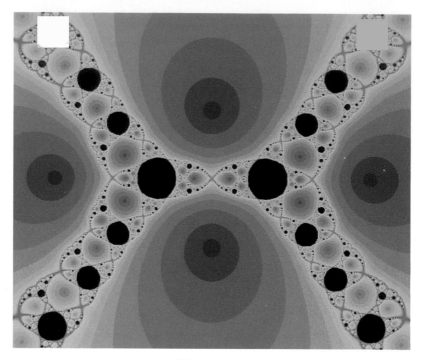

Fig. 18.5a

Fig. 18.5b

$F(z) = z^2 - 1$ converges to a root provided we choose an initial seed off the imaginary axis. If we choose the seed on the imaginary axis, then Newton's method does not converge to a root and leads, in fact, to chaotic behavior.

The following example explains something we discussed in detail in Chapter 13. Recall that, if we try to apply real Newton iteration to the function $F(x) = x^2 + 1$, which has no real roots, we observe chaotic behavior. This behavior can now be seen to be equivalent to the previous example.

Example. Let $F(z) = z^2 + 1$ with complex Newton iteration function

$$N(z) = \frac{1}{2}\left(z - \frac{1}{z}\right).$$

Let

$$H(z) = \frac{z - i}{z + i}.$$

Note that H takes the roots of F to zero and infinity. It is easy to verify that, as in the previous example, H is a conjugacy between N and $Q(z) = z^2$. See exercise 2.

The previous two examples suggest that, when Newton's method is applied to a complex quadratic polynomial with distinct roots at z_1 and z_2 in the plane,

1. N is chaotic on the line that is the perpendicular bisector of z_1 and z_2.
2. All other orbits of N converge to z_1 or z_2, depending upon which side of the perpendicular bisector the seed lies.

We ask you to verify these facts in exercise 3. Thus, as long as we choose points off the perpendicular bisector, Newton's method always converges in the quadratic case.

This fact was first proved by the British mathematician Cayley in 1879*. In a bit of mathematical overeagerness, Cayley wrote that he hoped that similar methods could be used to solve cubic polynomials. This proof, however, never appeared. There is a good reason for this: the cubic case is much more difficult than the quadratic case. In fact, if N has at least two distinct

---

*   Arthur Cayley, "On the Newton-Fourier Imaginary Problem," *Amer. J. Math* **2**, 97 (1879).

roots, then the set of points at which N fails to converge is a fractal, much like the Julia sets of Chapter 16. Moreover, it is entirely possible for N to admit attracting cycles that have period greater than 1. Orbits of points that lie in the basin of a cycle of high period therefore cannot be attracted to one of the roots, and so Newton's method fails to converge on a large set of points.

To plot the results of Newton's method in the complex plane, we make use of the following algorithm. Given a grid of points in the plane, and given a small $\epsilon > 0$, we will compute the orbit of each grid point under N for a maximum number of iterations, say K. If two successive points on the orbit ever lie within ϵ units of one another, then we will assume that the orbit has entered the vicinity of an attracting fixed point. More precisely, if $|N^j(z) - N^{j-1}(z)| < \epsilon$ and $N^j(z)$ is the first point on the orbit for which this is true, then we stop iterating and color the original point z with a color corresponding to j. We choose colors as before, with red indicating points whose orbits converge quickly, followed in order by orange, yellow, green, blue, indigo, and violet. If the orbit of z never satisfies the above condition during the first K iterations, then we leave z black.

Figure 18.3 plots the results of using this technique for Newton's method corresponding to $F(z) = z^2 + 1$ and $G(z) = z^3 - 1$. Note that the roots $\pm i$ of F and the cube roots of unity for G are plainly visible as the centers of the darkest red circles in each picture. In the case of F, the nonconvergent orbits lie along the real line, as we saw above. In the case of the cubic polynomial G, the nonconvergent points form a much more complicated subset of the plane. These sets are the analogs of the Julia sets (the chaotic regimes) for rational functions such as the Newton iteration functions.

Example. Let $F_a(z) = (z^2 - 1)(z^2 + a)$. The Newton iteration function for F_a is

$$N_a(z) = \frac{3z^4 + (a-1)z^2 + a}{4z^3 + 2(a-1)z}.$$

Recall from Chapter 13 that N_a has an attracting 2-cycle at the critical points

$$c_\pm = \pm\sqrt{\frac{1-a}{6}}.$$

where $a = 0.197017\ldots$. This cycle and all of the points attracted to it are colored black in Figure 18.4a. When the parameter a decreases, we begin to see some familiar behavior emerging. For example, when $a = 0.16$, the regions of nonconvergence for the corresponding Newton iteration are displayed in Figure 18.4b. Note that these regions resemble distorted versions of the filled Julia sets for the quadratic polynomial $Q_{-1}(z) = z^2 - 1$. In fact, one may show that this iteration admits an attracting cycle of period 4.

An interesting experiment arises when we ask for the fate of the orbits of the critical points $c_{\pm}$ as a function of the parameter a. Suppose we consider one of the two critical points, say c_+. We now write

$$c_+(a) = \sqrt{\frac{1-a}{6}}$$

to indicate the dependence of c_+ on a. As we have seen, the orbit of $c_+(a)$ may be attracted to one of the fixed points of N_a, or it may not. Thus there is a dichotomy and we ask: For which a-values does $c_+(a)$ tend to one of the roots of F_a, and for which a-values does this not occur?

Project. Design a computer program to plot the set of a-values for which $c_+(a)$ does not tend to one of the four roots of $F_a(z) = (z^2 - 1)(z^2 + a)$. What image do you see in the vicinity of $a = 0.16$?

The surprise appearance of a Mandelbrot set in these pictures is a phenomenon that has only recently begun to be understood. Some of the computer experiments that motivated this work were performed by Curry, Garnett, and Sullivan*, while the rigorous explanation was provided by Douady and Hubbard**.

Project. Consider the real polynomials

$$F_a(x) = (x^2 - 1)(x^2 + a).$$

Design a program to plot the orbit diagram for the associated Newton iteration function using $c_+(a)$ as the seed for each orbit. What do you now see in the vicinity of $a = 0.16$?

---

*   J. Curry, L. Garnett, and D. Sullivan, "On the iteration of a rational function: computer experiments with Newton's Method," *Comm. Math. Phys.* **91**, 267-277 (1983).

**   A. Douady and J. Hubbard, "On the dynamics of polynomial-like mappings," *Ann. Scient., Éc. Norm. Sup. 4ᵉ séries*, **18**, 287 (1985).

We encourage the reader to experiment with Newton's method for other complex polynomials. A program to effect the algorithm for Newton's method is contained in Appendix B. Much of the dynamics that we have touched upon in this book is present in various guises in these dynamical systems.

Exercises

1. Prove that the function $H(z) = \frac{z-1}{z+1}$ is one-to-one.

2. Show that the function $H(z) = \frac{z-i}{z+i}$ gives a conjugacy between the Newton iteration function for $F(z) = z^2 + 1$ and $Q(z) = z^2$.

3. Let $a, b \in \mathbf{C}$ with $a \neq b$. Consider the quadratic polynomial $F(z) = (z - a)(z - b)$. Show that the Newton iteration function corresponding to F is conjugate to $Q(z) = z^2$ via a conjugacy that takes a to zero and b to infinity.

4. Describe the dynamics of the Newton iteration function corresponding to $F(z) = z^d$ for $d = 2, 3, 4, \ldots$.

5. Describe the dynamics of the Newton iteration function corresponding to $E(z) = e^z$.

6. Let $F(z) = z^2(z - 1)$. Show that the Newton iteration function corresponding to F is conjugate to the logistic function $L(z) = \frac{1}{2}z(1 - z)$ via the conjugacy

$$H(z) = \frac{1 - z}{z}.$$

7. Prove that the Newton iteration function corresponding to the function $F(z) = ze^z$ is conjugate to the quadratic polynomial $L(z) = z + z^2$.

APPENDIX A

Mathematical Preliminaries

In this Appendix we will review some of the necessary notions from calculus that we will use in this book. We will also introduce a few elementary topological notions that we will use throughout. In general, readers may skip this section, referring back to it only if necessary. There will be certain notions that arise during the course of this book that are beyond the scope of typical calculus courses. We will in general describe these ideas completely when needed.

A.1 Functions

For the main part of this book, we will consider functions of one real variable. Later, when we introduce symbolic dynamics, we will encounter functions defined on other "spaces." Still later we will turn to functions of several variables and functions of a complex variable. For now, however, a function will be a real-valued function of one real variable, just as encountered in elementary calculus.

Recall that a function is an operation that assigns to certain real numbers x a new real number $F(x)$. The set of allowable x-values is called the *domain* of the function. The set of allowable values for $F(x)$ is called the *target space* or *range*. Often the domain of the function will be all real numbers, but occasionally we will restrict our attention to certain subsets of the reals. We will always denote a function by capital letters, as, for example, the cubing function F given by $F(x) = x^3$. Here F denotes the function and $F(x)$ denotes the value of the function at x, namely x^3. It is important to distinguish between the function—the operation of cubing—and its value at a particular point x, namely x^3.

We denote the set of all real numbers by $\mathbf{R}$. We write $F\colon \mathbf{R} \to \mathbf{R}$ to indicate that F is a real-valued function of a real variable. If the domain of F is the interval $[a, b]$ and we wish to specify the target as the interval $[c, d]$, we will write $F\colon [a, b] \to [c, d]$. This kind of notation will be important later when we discuss functions defined on the plane or other "spaces." It also emphasizes the fact that a function is a "dynamic" object that takes points in $[a, b]$ and assigns values to them in $[c, d]$. We think of F geometrically as moving or *mapping* points in the domain to points in the target. For this reason we will often refer to a function as a *map* or *mapping*. Figure A.1 displays the dynamic nature of functions.

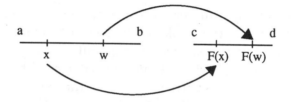

Fig. A.1 The function F maps points in $[a, b]$ to points in $[c, d]$.

Definition. A function F is called *one-to-one* if for any x_0 and x_1 in the domain of F with $x_0 \neq x_1$ we have $F(x_0) \neq F(x_1)$. We say that F is *onto* if for each y in the target space, there is an x in the domain of F such that $F(x) = y$.

Briefly, one-to-one functions assign different values to different numbers in the domain. Onto functions assume all values in the target. For example, increasing or decreasing functions are always one-to-one. They need not be onto, however. For example, if we consider the function $A(x) = \arctan(x)$ as a function $A\colon \mathbf{R} \to \mathbf{R}$, then A is increasing but not onto. If the target space is the open interval $-\pi/2 < x < \pi/2$, then this function is onto. We must always specify what the target space is when discussing whether or not a function is onto. As another example, the function $T\colon (-\pi/2, \pi/2) \to \mathbf{R}$ given by $T(x) = \tan(x)$ is both one-to-one and onto. Note that T is not one-to-one if we consider its domain to be all real numbers.

For a function F that is both one-to-one and onto, we may define the inverse function F^{-1}. This is the function that satisfies $F^{-1}(F(x)) = x$ for all x in the domain of F and $F(F^{-1}(x)) = x$ for all x in the domain of F^{-1}. For example, $A(x) = \arctan(x)$ is the inverse of $T(x) = \tan(x)$ when the domain of T is the interval $(-\pi/2, \pi/2)$. Note that the domain of A is the target of T while the target of A is the domain of T, so we could also write $A = T^{-1}$ or $T = A^{-1}$. As another example, the squaring function $S(x) = x^2$ is not one-to-one on all of $\mathbf{R}$, but this function is one-to-one on the interval $[0, \infty)$. So if we regard $S: [0, \infty) \to [0, \infty)$, then its inverse is $S^{-1}(x) = \sqrt{x}$ with $S^{-1}: [0, \infty) \to [0, \infty)$. We remark that the notation F^{-1} will always denote the inverse of F. This notation will never mean $1/F(x)$.

Given a point y in the target of F, there are often several points in the domain that are mapped to this point (there is only one if F is one-to-one). The *preimage* of y is the set of all points x that satisfy $F(x) = y$. We denote this set by $F^{-1}(y)$. Note the subtlety of notation here. The notation F^{-1} means the function that is the inverse of F. The notation $F^{-1}(y)$ is quite different: this is the set of preimages of y, and is defined even if F is not one-to-one. For example, the preimage of 0 under $F(x) = x^3 - x$ includes the three points 0, 1, and -1. So $F^{-1}(0) = \{0, \pm 1\}$. On the other hand, the preimage of 6 under F consists of only the point 2. This is true since we may write

$$x^3 - x - 6 = (x - 2)(x^2 + 2x + 3),$$

and the roots of $x^2 + 2x + 3$ are not real.

One of the most important notions involving functions is *continuity*. Often, in introductory calculus courses, this notion is introduced somewhat informally. For example, a function $F: \mathbf{R} \to \mathbf{R}$ is sometimes said to be continuous if one can draw its graph without lifting pen from paper. While this is an illustrative notion for functions of a real variable, this definition is not suitable for the kinds of functions that we will study (particularly in Chapter 9). Hence we need a more rigorous notion of continuity.

Definition. Let $F: X \to Y$ be a function. We say that F is *continuous at* x_0 if

    1. $F(x_0)$ is defined.

    2. $\lim_{x \to x_0} F(x) = F(x_0)$.

The function F is *continuous* if it is continuous at x_0 for all $x_0 \in X$.

Example. The doubling function D: $[0,1) \to [0,1)$

$$D(x) = \begin{cases} 2x & 0 \le x < 1/2 \\ 2x - 1 & 1/2 \le x < 1 \end{cases}$$

is not continuous at $x = 1/2$ since $D(1/2) = 0$, but the limit of $D(x)$ as x approaches $1/2$ from the left is 1. However, D is continuous at all $x \ne 1/2$ in the domain.

Example. The function $F(x) = 1/x$ is not continuous at 0 since $F(0)$ is not defined. However, F is a continuous function on the domain $0 < |x| < \infty$.

Definition. A function F is a *homeomorphism* if it has the following properties:

1. F is one-to-one.
2. F is onto.
3. F is continuous.
4. F^{-1} is also continuous.

If F is a homeomorphism, we say that the domain of F and its target space are *homeomorphic*. For example, $F(x) = x^{1/3}$ is a homeomorphism that takes the reals to the reals with inverse $F^{-1}(x) = x^3$. The function $A : \mathbf{R} \to (-\pi/2, \pi/2)$ given by $A(x) = \arctan(x)$ is also a homeomorphism with inverse $T(x) = \tan(x)$. This shows that the interval $(-\pi/2, \pi/2)$ is homeomorphic to the entire real line.

A.2   Some Ideas from Calculus

Most of the results in this book depend upon a solid knowledge of differential calculus in one variable. Later we will talk about derivatives of complex functions, but these ideas will be introduced as needed.

Recall from calculus that the derivative of a function F at x_0 is defined by

$$F'(x_0) = \lim_{h \to 0} \frac{F(x_0 + h) - F(x_0)}{h}$$

or, equivalently,

$$F'(x_0) = \lim_{x \to x_0} \frac{F(x) - F(x_0)}{x - x_0}.$$

A function is said to be *differentiable at* x_0 if this limit exists. F is said to be *differentiable* if it is differentiable at all x_0 in the domain of F. Recall that $F'(x_0)$ gives the slope of the tangent line to the graph of F at $(x_0, F(x_0))$.

We will make use of the higher derivatives $F''(x_0)$ and $F'''(x_0)$ at various points in this book. Thus we will tacitly assume that all derivatives of F exist, unless otherwise noted. Such functions are called *smooth* or C^∞. Occasionally, we will introduce a nondifferentiable or noncontinuous example, such as the doubling function.

Example. The function $F(x) = |x|$ is not differentiable at $x_0 = 0$, since

$$\lim_{x \to 0^+} \frac{F(x) - F(0)}{x} = 1$$

$$\lim_{x \to 0^-} \frac{F(x) - F(0)}{x} = -1.$$

Also, $G(x) = x^{1/3}$ is not differentiable at $x = 0$ since

$$\lim_{x \to 0} \frac{G(x) - G(0)}{x} = \infty.$$

These examples represent two of the fundamental ways a function may fail to be differentiable at a point.

One important fact from calculus is

Differentiability $\Rightarrow$ Continuity. *If $F'(x_0)$ exists, then F is continuous at* x_0.

Example. The doubling function $D\colon [0, 1) \to [0, 1)$

$$D(x) = \begin{cases} 2x & 0 \le x < 1/2 \\ 2x - 1 & 1/2 \le x < 1 \end{cases}$$

is not differentiable at $x = 1/2$ since it is not continuous at this point. On the other hand, $D'(x) = 2$ for all $x \in [0, 1)$, $x \ne 1/2$.

There are a number of other important theorems in calculus that we will use, including the Intermediate Value Theorem and the Mean Value Theorem. We will remind you of these results in Chapter 5 when we first need them.

A.3   Open and Closed Sets

We will often consider functions defined on closed or open intervals. A closed interval is any interval of the form $a \leq x \leq b$, or $[a, b]$ for short. According to this definition, the single point $\{a\}$ is a closed interval. Similarly, an open interval is any interval of the form $a < x < b$, or (a, b) for short.

So that we don't get into logical difficulties below, we will declare that any semi-infinite interval of the forms $[a, \infty)$ or $(-\infty, a]$ are closed, while the intervals (a, ∞) or $(-\infty, a)$ are open. This means that the entire real line is to be regarded as both a closed and an open interval. In particular, the empty set, the complement of the entire real line, is also both a closed and an open interval, by definition.

Note that the intersection of two open intervals is another open interval. Also, the intersection of two closed intervals is a closed interval. Even if either of these intervals is empty, our convention above guarantees that these results are true.

A subset $X \subset \mathbf{R}$ is an open set if, for any point $x \in X$, we can find a small open interval about x that lies entirely in X. A subset $\overline{Y} \in \mathbf{R}$ is said to be closed if its complement in $\mathbf{R}$ is open. Remember that the complement of a subset $\overline{Y}$ in $\mathbf{R}$ consists of all points in $\mathbf{R}$ that are not in $\overline{Y}$. Subsets of $\mathbf{R}$ may be neither open nor closed, as for example the interval $(a, b]$.

One important fact about open and closed sets is the following. If A_1 and A_2 are two open sets in $\mathbf{R}$, then so too is $A_1 \cup A_2$, since about any point in $A_1 \cup A_2$, we may always find an open interval in either A_1 or A_2 so this interval lies in $A_1 \cup A_2$. Similarly, $A_1 \cap A_2$ is also an open set. Likewise, if B_1 and B_2 are closed sets, then $B_1 \cap B_2$ and $B_1 \cup B_2$ are also closed. (This is again why we need $\mathbf{R}$ and the empty set to be both closed and open.)

Instead of taking the union or intersection of two or, more generally, finitely many open or closed sets, we will occasionally have to take unions or intersections of countably* many intervals. Here things are slightly more complicated.

Countable Union Property.   *If A_i is an open set for $i = 1, 2, \ldots$ then $\bigcup_{i=1}^{\infty} A_i$ is also an open set. If B_i is a closed set, then $\bigcup_{i=1}^{\infty} B_i$ need not be a closed set.*

---

*   A collection of sets is countable if it can be put in one-to-one correspondence with the natural numbers.

In fact, the union of any collection of open sets is open, for if x lies in this union, then x must lie in some A_i, and so there is an open interval about x in A_i that thus lies in the union. The following example shows that countable unions of closed set need not be closed. Let $B_n = [\frac{1}{n}, 1]$ for $n = 1, 2, \ldots$. Then

$$\bigcup_{n=1}^{\infty} B_n = (0, 1],$$

which is not closed.

Countable Intersection Property. *If B_i, $i = 1, 2, \ldots$, are closed sets, then $\bigcap_{i=1}^{\infty} B_i$ is closed. If A_i, $i = 1, 2, \ldots$, are open sets, then $\bigcap_{i=1}^{\infty} A_i$ need not be open.*

Recall that the complement of a closed set is open and note that the complement of the intersection of countably many closed sets is the union of countably many open sets. Since this union is open, it follows that the intersection of countably many closed sets is closed. An example of the second statement is provided by the open intervals $A_n = (-1/n, 1/n)$. Note that

$$\bigcap_{n=1}^{\infty} A_n = \{0\},$$

which is a closed set.

Two closed intervals A and B are said to be nested if $A \subset B$. Our next observation in this section is:

Nested Intersection Property. *Suppose B_i, $i = 1, 2, \ldots$ are closed, nonempty intervals. Suppose $B_1 \supset B_2 \supset \ldots$. Then $\bigcap_{i=1}^{\infty} B_i$ is nonempty.*

The reason for this is simple. Let $B_i = [l_i, r_i]$. Then our assumption is that

$$l_1 \le l_2 \le l_3 \le \cdots \le r_3 \le r_2 \le r_1.$$

Let $l = \lim_{i \to \infty}$ and $r = \lim_{i \to \infty} r_i$. Clearly, $l \le r$. Then the interval $[l, r]$ lies in the intersection of all of the B_i. Note that we may have $l = r$ so that $\bigcap_{i=1}^{\infty} B_i = \{l\}$, which is still nonempty.

We observe that if I is an open interval in $\mathbf{R}$ and $F : \mathbf{R} \to \mathbf{R}$ is continuous, then $F^{-1}(I)$ is an open subset of $\mathbf{R}$. To see why this is true, suppose that $F^{-1}(I)$ contains an interval of the form $[a, b]$, but that (b, c) does not lie in

$F^{-1}(I)$ for some $c > b$. Since $F(b)$ lies in I and I is open, there is some small interval surrounding $F(b)$ that lies entirely in I. If F takes the interval (b, c) outside of I, then F is necessarily discontinuous at b. This contradiction establishes the result.

Arguing similarly using complements, it is easy to see that if I is a closed interval, then $F^{-1}(I)$ is a closed subset of $\mathbf{R}$ if $F: \mathbf{R} \to \mathbf{R}$ is continuous. Thus:

Preimage Property. *Suppose $F: \mathbf{R} \to \mathbf{R}$ is continuous. If I is an open interval in $\mathbf{R}$, then $F^{-1}(I)$ is an open set. If I is a closed interval, then $F^{-1}(I)$ is a closed set.*

Another important notion is that of *connectedness*. A subset S of the real line is connected if it has the property that whenever $a, b \in S$ and $a \leq c \leq b$, then c is also in S. That is, if S contains any two points a and b, then S contains all points in between. Equivalently, S is a single interval, which may be open, closed or half-open, half-closed. In particular, our remarks after the Nested Intersection Property show that the nested intersection of closed, connected sets is also connected.

Connectedness Property. *Suppose B_i, $i = 1, 2, \ldots$ are closed, nonempty intervals. Suppose $B_1 \supset B_2 \supset \ldots$. Then $\bigcap_{i=1}^{\infty} B_i$ is a connected set, in fact, a closed interval.*

We will also need to deal with closed and open subsets of the plane. A subset S of the plane is *open* if, for each $z_0 \in S$, we may find a small disk of the form $|z - z_0| < \epsilon$ that is entirely contained in S. A subset of the plane is *closed* if its complement is open. By definition, the entire plane is both a closed and an open set. For our purposes, we need only consider the notion of closed, connected subsets of the plane. A closed set in the plane is said to be *connected* if it cannot be written as the union of two disjoint closed subsets of the plane. While we will not go into the proofs here, it is a fact that all five of the above properties hold for closed and open subsets of the plane (replacing the term closed or connected interval with closed or connected set).

APPENDIX B

Algorithms

In this appendix we include several True BASIC programs that will enable the user to display some of the images found in this text. These programs are by no means optimal in terms of both runtime or usability. Rather, they should serve mainly as guides for the reader who wishes to develop custom-made software for demonstration and/or experimentation purposes. As the reader will soon discover, many of these programs take a long time to run. We recommend that serious users consider converting the code to to other languages such as C or PASCAL in order to accelerate the computation time.

Iteration is a natural process to carry out on a computer. There is nothing that a computer can do faster than iterate a simple function. Moreover, as the following program illustrates, the programs that accomplish iteration are quite short and easy to write.

```
!
!   iterate
!
INPUT PROMPT ''Parameter?  '':  c
INPUT PROMPT ''Maximum number of iterations?  '':  maxiter
INPUT PROMPT ''Initial seed?  '':  x
PRINT 0, x
FOR i = 1 TO maxiter
    LET x = x*x + c
    PRINT i,x
NEXT i
END
```

Spreadsheets are also an excellent tool with which to demonstrate the process of iteration. The following spreadsheet prompts the user to select a c value and a seed x_0 and then computes the first 25 points on the orbit of x_0 under iteration of $Q_c(x) = x^2 + c$. We display the appropriate formulas in columns A and B, and the output in columns C and D.

	A	B
1	Parameter:	Seed:
2	- 2	0.2
3		
4		Orbit:
5	0	=B2
6	=A5+1	=B5*B5+$A$2
7	=A6+1	=B6*B6+$A$2
8	=A7+1	=B7*B7+$A$2
9	=A8+1	=B8*B8+$A$2
10	=A9+1	=B9*B9+$A$2
11	=A10+1	=B10*B10+$A$2
12	=A11+1	=B11*B11+$A$2
13	=A12+1	=B12*B12+$A$2
14	=A13+1	=B13*B13+$A$2
15	=A14+1	=B14*B14+$A$2
16	=A15+1	=B15*B15+$A$2
17	=A16+1	=B16*B16+$A$2
18	=A17+1	=B17*B17+$A$2
19	=A18+1	=B18*B18+$A$2
20	=A19+1	=B19*B19+$A$2
21	=A20+1	=B20*B20+$A$2
22	=A21+1	=B21*B21+$A$2
23	=A22+1	=B22*B22+$A$2
24	=A23+1	=B23*B23+$A$2
25	=A24+1	=B24*B24+$A$2
26	=A25+1	=B25*B25+$A$2
27	=A26+1	=B26*B26+$A$2
28	=A27+1	=B27*B27+$A$2
29	=A28+1	=B28*B28+$A$2
30	=A29+1	=B29*B29+$A$2

	C	D
1	Parameter:	Seed:
2	- 2	0.2
3		
4		Orbit:
5	0	0.2
6	1	-1.96
7	2	1.8416
8	3	1.39149056
9	4	-0.063754021
10	5	-1.995935425
11	6	1.98375822
12	7	1.935296675
13	8	1.745373218
14	9	1.046327672
15	10	-0.905198403
16	11	-1.180615851
17	12	-0.606146213
18	13	-1.632586768
19	14	0.665339555
20	15	-1.557323277
21	16	0.425255788
22	17	-1.819157515
23	18	1.309334064
24	19	-0.285644309
25	20	-1.918407329
26	21	1.680286679
27	22	0.823363322
28	23	-1.32207284
29	24	-0.252123407
30	25	-1.936433788

A spreadsheet to compute orbits of the quadratic function

Obviously, the spreadsheet and the preceding program may easily be modified to compute orbits for other functions. For example, using the function fp(x) that gives the fractional part of x, we may rewrite the program to iterate the doubling function.

```
!
!  doubling-function
!
INPUT PROMPT ''Maximum number of iterations?  '':  maxiter
INPUT PROMPT ''Initial seed?  '':  x
PRINT 0, x
FOR i = 1 TO maxiter
    LET x = fp(2*x)
    PRINT i,x
NEXT i
END
```

Graphical analysis is one of the most dynamic methods of displaying the behavior of orbits. The following program first plots the graph of the quadratic function $Q_c(x) = x^2 - 2$ over the interval $-2 \le x \le 2$. Then the user selects an initial seed in this interval and the program displays the first 50 points on the orbit of this point via graphical analysis.

```
!
!  graphical-analysis
!
OPEN #1:  SCREEN .25, .75, .1, .8
SET WINDOW -2, 2, -2, 2
!  draw axes and diagonal
PLOT -2, -2; 2, 2
PLOT 0, -2; 0, 2
PLOT -2, 0; 2, 0
!  sketch the graph
FOR x = -2 TO 2 STEP .005
    PLOT x, x*x - 2
NEXT x
GET POINT x, t
LET x0 = x
```

```
SET COLOR ''red''
FOR i = 1 TO 50
     LET y = x*x - 2
     PLOT x, x; x, y; y, y
     LET x = y
NEXT i
END
```

Recall that the orbit diagram for the quadratic function gives a picture of the asymptotic orbit of 0 for a variety of c-values in the interval $-2 \le c \le 0.25$. In the following program we will plot 400 equally spaced c-values horizontally. Over each such c, we will compute the first 200 points on the orbit of 0 but display only the last 150 of these points.

```
!
!  orbit-diagram
!
SET WINDOW -2, .25, -2, 2
FOR c = .25 TO -2 STEP -0.005625
     LET x = 0
     FOR i = 0 TO 200
          LET x = x*x + c
          IF i > 50 THEN PLOT c, x
     NEXT i
NEXT c
END
```

The programs to compute the Mandelbrot set and Julia sets are easy to write but take a long time to run. The following program produces an image of the Mandelbrot set in a square region in the complex plane. The program prompts the user to select the square by specifying the coordinates of the lower left vertex and the length of a side. We then consider a grid of 400 by 400 equally spaced points in the square. Each grid point represents a parameter value c. So, for each such point, the program then computes the orbit of c under Q_c, at least for the first `maxiter` iterations. If a point on the orbit ever has modulus larger than 2, then we know from the Algorithm in Chapter 17 that the orbit tends to ∞, so we color the original c-value white. On the other hand, if all points on the orbit remain on or inside the circle of radius 2, then we assume that the original point is in the Mandelbrot set and color this point black. Of course, since we are only using a finite number of iterations to make this determination, we may have erred in deciding that this point is in the Mandelbrot set. Consequently, the black points generated by this program yield only an approximation to the Mandelbrot set.

We caution the reader to use relatively few iterations when using this program. For large squares in the plane, no more than 50 iterations should be used. For more detailed images in smaller regions, the number of iterations will have to be raised, however. Also, recall that we know the fate of any c with $|c| > 2$. Therefore it makes sense only to use this algorithm in squares inside the square of side length 4 centered at the origin.

```
!
!  mandelbrot
!
INPUT PROMPT ''lower left coordinates?  '':  a1,b1
INPUT PROMPT ''sidelength?  '':  s
INPUT PROMPT ''maximum number of iterations?  '':  maxiter
LET a2 = a1 + s
LET b2 = b1 + s
OPEN #1:  SCREEN .25, .75, .1, .8
SET WINDOW a1, a2, b1, b2
LET size = s/400
FOR c1 = a1 TO a2 STEP size
  FOR c2 = b1 TO b2 STEP size
    LET x = 0
    LET y = 0
    LET i = 0
    DO WHILE i < maxiter
      LET r = x*x + y*y
      IF r > 4 THEN EXIT DO
      LET t = x*x - y*y + c1
      LET y = 2*x*y + c2
      LET x = t
      LET i = i + 1
    LOOP
      IF i = maxiter THEN PLOT c1, c2
  NEXT c2
NEXT c1
END
```

It is easy to modify the program that generates the Mandelbrot set to produce Julia sets instead. We begin by fixing a certain parameter $c = c_1 + ic_2$. As before, we select a square region in the complex plane. Then, instead of treating each point in the grid as a different parameter value, we now consider each grid point as the seed of an orbit for Q_c. By the corollaries to the Escape Criterion of Chapter 16, we need only check whether the orbit ever enters the region outside of the circle of radius 2 centered at the origin, provided $|c| \leq 2$. This is accomplished in much the same manner as in the program for the Mandelbrot set.

```
!
!  julia
!
INPUT PROMPT ''real part of c?  '':  c1
INPUT PROMPT ''imaginary part of c?  '':  c2
INPUT PROMPT ''lower left coordinates?  '':  a1,b1
INPUT PROMPT ''sidelength?  '':  s
INPUT PROMPT ''maximum number of iterations?  '':  maxiter
LET a2 = a1 + s
LET b2 = b1 + s
OPEN #1:  SCREEN .25, .75, .1, .8
SET WINDOW a1, a2, b1, b2
LET size = s/400
FOR x0 = a1 TO a2 STEP size
   FOR y0 = b1 TO b2 STEP size
      LET x = x0
      LET y = y0
      LET i = 0
      DO WHILE i < maxiter
         LET r = x*x + y*y
         IF r > 4 THEN EXIT DO
         LET t = x*x - y*y + c1
         LET y = 2*x*y + c2
         LET x = t
         LET i = i + 1
      LOOP
         IF i = maxiter THEN PLOT x0, y0
   NEXT y0
NEXT x0
END
```

Finally, we present an algorithm that will produce a picture of the convergence of Newton's method in the complex plane. We will describe a program that works for the Newton iteration function corresponding to $F(z) = z^3 - 1$, i.e.,

$$N(z) = \frac{2z^3 + 1}{3z^2}.$$

We will again select a square in the complex plane. For each point in a 400 by 400 grid in this square, we will compute the corresponding orbit under N. We stop iterating whenever we come within `epsilon` units of one of the three roots of F, or whenever we reach the maximum number of iterations, `maxiter`. Unlike the plates in Chapter 18, we will then color our initial seed according to which root the Newton iteration has converged. For example, if the orbit tends toward the root at 1, we color the initial point red. If it converges to the root in the upper (resp. lower) half plane, we color the initial point green (resp. blue). If the orbit does not converge within `maxiter` iterations, we color the seed black. This method has the advantage of showing clearly the basins of attraction of the different roots. Note that the program needs to be substantially rewritten to find the roots of other functions.

```
!
!  newton's method in 3 colors
!
INPUT PROMPT ''lower left coordinates?  '':  a1,b1
INPUT PROMPT ''sidelength?  '':  s
INPUT PROMPT ''maximum number of iterations?  '':  maxiter
LET a2 = a1 + s
LET b2 = b1 + s
OPEN #1:  SCREEN .25, .75, .1, .8
SET WINDOW a1, a2, b1, b2
LET size = s/400
LET epsilon = .01
FOR x0 = a1 TO a2 STEP size
   FOR y0 = b1 TO b2 STEP size
      LET x = x0
      LET y = y0
      LET i = 0
      DO WHILE i < maxiter
         LET z = x*x*x*x + y*y*y*y +2*x*x*y*y
         IF z > 100000 THEN EXIT DO
```

```
      LET x1 = (2*x + (x*x - y*y)/z)/3
      LET y1 = (2*y -2*x*y/z)/3
      IF ABS(x-x1) < epsilon AND ABS(y-y1) < epsilon THEN
        IF x1 > 0 THEN
           SET COLOR ''red''
           PLOT x0, y0
           EXIT DO
        ELSE IF y1 > 0 THEN
           SET COLOR ''green''
           PLOT x0, y0
           EXIT DO
        ELSE y1 < 0 THEN
           SET COLOR ''blue''
           PLOT x0, y0
           EXIT DO
        END IF
      END IF
      LET x = x1
      LET y = y1
      LET i = i + 1
      IF i = maxiter THEN PLOT x0, y0
   LOOP
  NEXT y0
NEXT x0
END
```

APPENDIX C

References

There are now many references available for students who wish to pursue further study of the topics in this book. The following books begin where this text leaves off. Each treats the material at a more advanced level.

Arnol'd, V. *Mathematical Methods of Classical Mechanics,* Springer-Verlag, New York, 1978.

Arnol'd, V. *Geometrical Methods in the Theory of Ordinary Differential Equations,* Springer-Verlag, New York, 1983.

Baker, G. L. and Gollub, J. P. *Chaotic Dynamics: An Introduction,* Cambridge University Press, Cambridge, 1990.

Beardon, A. *Iteration of Rational Functions,* Springer-Verlag, New York, 1991.

Collet, P. and Eckmann, J.-P. *Iterated Maps on the Interval as Dynamical Systems,* Birkhäuser, Boston, 1980.

Devaney, R. L. *An Introduction to Chaotic Dynamical Systems,* Second Edition. Addison-Wesley, Redwood City, Calif., 1989.

Devaney, R. L. and Keen, L., eds. *Chaos and Fractals: The Mathematics Behind the Computer Graphics,* American Mathematical Society, Providence, 1989.

Guckenheimer, J. and Holmes, P. *Nonlinear Oscillations, Dynamical Systems, and Bifurcations of Vector Fields,* Springer-Verlag, New York, 1983.

Hirsch, M. W. and Smale, S. *Differential Equations, Dynamical Systems, and Linear Algebra,* Academic Press, New York, 1974.

Milnor, J. *Dynamics in One Complex Variable: Introductory Lectures,* SUNY Stony Brook, Institute for Mathematical Sciences, Preprint #1990/5, 1990.

Palis, J. and deMelo, W. *Geometric Theory of Dynamical Systems,* Springer-Verlag, New York, 1982.

Ruelle, D. *Elements of Differentiable Dynamics and Bifurcation Theory,* Academic Press, San Diego, 1989.

Shub, M. *Global Stability of Dynamical Systems,* Springer-Verlag, New York, 1986.

Tufillaro, N., Abbott, T., and Reilly, J. *An Experimental Approach to Nonlinear Dynamical Systems,* Addison-Wesley, Redwood City, CA, 1992.

Wiggins, S. *Global Bifurcations and Chaos,* Springer-Verlag, New York, 1988.

For more details on the computer programs and algorithms used in this book, we recommend

Georges, J., Johnson, D. and Devaney R. L. *A First Course in Chaotic Dynamical Systems: Laboratory,* Addison-Wesley Co., Reading, MA, 1992.

Parker, T. and Chua, L. *Practical Numerical Algorithms for Chaotic Systems,* Springer-Verlag, New York, 1989.

Peitgen, H.-O. and Richter, P. *The Beauty of Fractals,* Springer-Verlag, Heidelberg, 1986.

Peitgen, H.-O. and Sajupe, D. *The Science of Fractal Images,* New York, Springer-Verlag, 1989.

The following articles are research papers. They are included here mainly for historical interest. Of course, there are many other such papers in the literature. These have been chosen since they relate to some of the topics covered in this book.

Fatou, P. "Sur les Équationes Fonctionelles." *Bull. Soc. Math. France* **48** (1920), 33-94, 208-314.

Feigenbaum, M. J. "Quantitative Universality for a Class of Nonlinear Transformations." *J. Stat. Phys.* **19** (1978), 25-52.

Julia, G. "Mémoire sur l'Itération des Fonctions Rationelles." *J. Math. Pures Appl.* **4** (1918), 47-245.

Li, T.-y., and Yorke, J., "Period Three Implies Chaos." *American Mathematical Monthly* **82** (1975), 985-992.

Lorenz, E. N. "Deterministic Nonperiodic Flows." *J. Atmospheric Sci.* **20** (1963), 130-141.

Mandelbrot, B. "Fractal Aspects of the Iteration of $z \mapsto \lambda z(1-z)$." *Nonlinear Dynamics,* Annals of the New York Academy of Science **357** (1980), 249-259.

Sarkovskii, A. N. "Coexistence of Cycles of a Continuous Map of a Line into Itself." *Ukrain. Mat. Z.* **16** (1964), 61-71 (In Russian).

Smale, S. " Diffeomorphisms with Many Periodic Points." *Differential and Combinatorial Topology,* Princeton Univ. Press, (1964), 63-80.

The following books represent a variety of popular treatments of chaos and fractals. They make wonderful supplementary reading for students taking a course in dynamics.

Gleick, J. *Chaos: Making a New Science,* Viking, New York, 1987.

McGuire. M. *An Eye for Fractals,* Addison-Wesley, Redwood City, CA, 1991.

Peterson, I. *The Mathematical Tourist,* W. H. Freeman, New York, 1988.

Schroeder, M. *Fractals, Chaos, Power Laws,* W. H. Freeman, New York, 1991.

The following books delve more deeply into the fractal geometry. Most discuss dynamical systems only peripherally.

Barnsley, M. *Fractals Everywhere,* Boston, Academic Press, 1988.

Edgar, G. A. *Measure, Topology, and Fractal Geometry,* Springer-Verlag, New York, 1990.

Falconer, K. *Fractal Geometry,* Wiley, Chichester, 1990.

Falconer, K. *The Geometry of Fractal Sets,* Cambridge University Press, Cambridge, 1985.

Kaye, B. *A Random Walk Through Fractal Dimensions,* VCH Publishers, New York, 1989.

Mandelbrot, B. *The Fractal Geometry of Nature,* W. H. Freeman, New York, 1983.

Prusinkiewicz, P. and Lindenmayer, A. *The Algorithmic Beauty of Plants,* Springer-Verlag, New York, 1990.

The following are lower level textbooks. They attempt to explain dynamical systems theory to students who do not have the benefit of a knowledge of calculus.

Abraham, R. and Shaw, C. *Dynamics: The Geometry of Behavior,* Addison-Wesley, Redwood City, CA, 1992.

Devaney, R. L. *Chaos, Fractals, and Dynamics: Computer Experiments in Mathematics,* Addison-Wesley, Menlo Park, 1990.

Tuck, E. O. and de Mestre, N. J. *Computer Ecology and Chaos,* Longman Cheshire, Melbourne, 1991.

One lower level text that emphasizes applications of dynamical systems theory is:

Sandefur, J. *Discrete Dynamical Systems: Theory and Applications,* Clarendon Press, Oxford, 1990.

INDEX